RUNNING SMART

RUNNING SMART

HOW SCIENCE CAN IMPROVE YOUR ENDURANCE AND PERFORMANCE

MARISKA VAN SPRUNDEL

TRANSLATED BY DANNY GUINAN

THE MIT PRESS   CAMBRIDGE, MASSACHUSETTS   LONDON, ENGLAND

This publication has been made possible with financial support from the Dutch Foundation for Literature.

N ederlands
letterenfonds
dutch foundation
for literature

This book was set in ITC Stone and Avenir by New Best-set Typesetters Ltd. Printed and bound in the United States of America.

Library of Congress Cataloging-in-Publication Data

Names: Sprundel, Mariska van, author.
Title: Running smart : how science can improve your endurance
    and performance / Mariska van Sprundel ; translated by Danny
    Guinan.
Other titles: Alles wat je wilt weten over hardlopen. English
Description: Cambridge, Massachusetts : The MIT Press, 2021. |
    Includes bibliographical references and index.
Identifiers: LCCN 2020029795 | ISBN 9780262542449 (paperback)
Subjects: LCSH: Running—Training. | Running—Physiological
    aspects. | Running races. | Endurance
Classification: LCC GV1061.5 .S6713 2021 | DDC 796.42071—dc23
LC record available at https://lccn.loc.gov/2020029795

10 9 8 7 6 5 4 3 2 1

CONTENTS

PREFACE vii

1   THE RISE OF THE LONG-DISTANCE RUNNER   1

2   THE USEFULNESS OF RUNNING SHOES   19

3   BACK TO BARE FEET   41

4   BUILT TO RUN   67

5   TRAINING LOAD AND LOAD CAPACITY   93

6   THE RIGHT FUEL IN THE TANK   123

7   A SPRINT TO THE JOHN   149

8   RUNNING FOR YOUR LIFE   167

9   THE SECRET TO SPEED   189

10 FATIGUE IS ALL IN THE MIND   209

11 RUNNING AS THERAPY FOR THE BRAIN   235

EPILOGUE: DO YOUR OWN SCIENCE   259
ACKNOWLEDGMENTS   265
BIBLIOGRAPHY   267
INDEX   291

PREFACE

No amount of belief makes something a fact.
—James Randi, pseudoscience skeptic

In the spring of 2016, I found myself standing on a beach among a crowd of spectators cheering on 17,000 runners during the Egmond half marathon, including a group of runners from my own running club in Utrecht in the Netherlands. A few weeks earlier I had come down with a shin injury, not realizing at the time that it would literally put me "out of the running" for a couple of months. I was bitterly disappointed. Egmond is the highlight of the year at our club and I wasn't taking part. Instead I was standing on a beach in the freezing cold, doing my best to muster some enthusiasm for my colleagues. I didn't know whether to laugh or cry. But cheering your friends on is better than staying home and sulking, of course. You get caught up in the excitement of the event and want to be there for your fellow runners. At the same time, I was very jealous. Everyone was running except me. Still, there was some cold comfort in the fact that the 2016 race would be remembered as one of the most difficult in living memory because of the terrible weather conditions. Everywhere I looked, I saw runners grimacing from the effort of trying to cope with the ferocious head wind and the sand blasting their faces. It didn't look like much fun, to be honest. In fact, I couldn't have picked a better year to get injured.

As the runners passed by, I observed the various techniques on display. Those out in front ran a lot differently from those at the rear ambling along just ahead of the safety car. The runners at the back of the field also wore warmer clothing. Some listened to music as they ran, others kept looking at their watches. Some waved breezily at the crowd cheering them on, while others were the picture of focus and concentration. Running gels were being consumed in abundance. One thing that intrigued me was the incredible variety of footwear. Everything from simple sneakers with thin soles to hi-tech running shoes in blindingly florescent colors. Every now and then someone ran by wearing so-called FiveFingers shoes, with individual pockets for the toes. Some wore no shoes at all, their bare feet sinking deep into the ice-cold sand. Chilled to the bone, I watched the colorful parade of runners as they passed by, and it was there, on the beach in Egmond, that the idea of writing this book was born. The world of running is awash with assumptions, rituals, and stories related to nutrition, shoes, mental strength, and injuries. The stories are generally passed on from runner to runner, and almost everyone has the same store of knowledge and knows what's good for you and what's not. Up until 2016 I had not given any of this much thought. I only started running in my twenties and had always accepted traditional running wisdom more or less as fact.

The science journalist in me sprang into action that year at the half marathon in Egmond. The first question that came to mind was: is everything we hear about running really true? I had just suffered another in a long line of injuries, even though I believed I was doing everything by the book. This got me thinking: Can good running shoes prevent injury? Is it better to run with a higher stride frequency or to take longer strides? Does beetroot juice boost your performance? When I recently became a trainer at my running club, I started paying more attention to what works and what doesn't. This sparked a personal crusade to discover "the science behind running." My questions led me to interview experts across a broad range of disciplines. I had my own running technique examined in an exercise lab and immersed myself in the scientific literature. *Running Smart* contains many insights from the world of sports science that will interest both recreational and professional runners. Some chapters offer practical advice, while others address the relationship between running and the human body. For example, is every kind of body suited to running? How often do you need to run in order to stay healthy? And how does jogging help to alleviate stress?

I wrote the bulk of this book in Boston and Cambridge, Massachusetts, where I spent six months working in 2016 and 2017. Boston is a mecca for runners. The historical city is famous for hosting the world's oldest annual marathon, which is also one of the Abbott World Marathon Majors (the other five being Tokyo, London, Berlin, Chicago, and New York City). My dream is to run the Boston Marathon someday. Boston and Cambridge are great places not only for running but also for science and technology, and I made good use of the fact that Harvard University and Harvard Medical School were only a ten-minute bike ride from where I lived. This made it a lot easier for me to visit people like evolutionary biologist Daniel Lieberman, who believes that humans were born to run, and biomechanical engineer Irene Davis, who examines running techniques in her exercise lab. Their research is addressed extensively in this book. One thing I quickly discovered is that there is often a discrepancy between the facts and accepted wisdom. Athletics is a minefield of ritual and tradition, and running is no exception.

While reading this book, it may be useful to keep in mind that scientific research is usually based on groups and not on individuals. The individual runner, with all their personal and physical traits, is not the same person as the average test subject in a study. As a result, you may wonder just how applicable some scientific findings are in real life. Treating injuries, combatting depression, finding the right running shoes, and preventing training overload all demand a made-to-measure approach to the individual runner and their situation. Nevertheless, there is still plenty of knowledge, for example with regard to running techniques and nutrition, that applies to almost everyone.

I am under no illusion that what I have discovered along the way will keep me injury-free for the rest of my running life. However, the next time I'm at a party I will be able to be the smarty-pants thanks to my newly acquired arsenal of information on shoes and stretching exercises. And pretty soon, so will you.

1

THE RISE OF THE LONG-DISTANCE RUNNER

==

The real mutants are the runners who don't get injured.
—Christopher McDougall, *Born to Run*

Before 1960, you would have been regarded as a bit of an oddball if you were fond of going for a jog in the park. All that huffing and puffing did nothing but disturb the peace, and, in the eyes of the general public, people who ran around town sporting skimpy shorts were simply looking for attention. The only acceptable reason for running in public was when you were trying to catch the bus. All this began to change, however, in the 1960s and '70s, when running moved from the relatively secluded arena of athletic clubs and university campuses to the greater public realm. The story of the rise of running as a recreational activity and the multimillion-dollar business it has spawned is told in the book *Running across Europe*, coauthored by Koen Breedveld from Radboud University in the Netherlands. The book is filled with mind-boggling statistics. Did you know, for example, that 826 out of every 1 million people in Europe have completed a marathon? Or that, in Europe, 50 million runners spend a total of $10.6 billion each year on their chosen sport, including on running shoes, clothing, and races? Finland tops the charts with an outlay of $910 per runner per annum. In the United States, a quarter of all runners questioned in a recent survey said that they spent between $101 and $120 on their last pair of running shoes. And the differences across the board in the amounts spent are enormous: 2 percent of those surveyed spent over $200 on their shoes, while the 2 percent at the other end of the scale bought running shoes that cost between $21 and $40.

Running first became widely popular in the United States, where jogging, as it was first called, became a socially acceptable activity. In addition, science began to produce more and more evidence of the importance of physical exercise to maintaining good health, which encouraged people to start jogging to stay fit. In the late 1970s, the jogging craze crossed the Atlantic to Europe, where the number of joggers grew steadily year after year. Scientists and researchers refer to this period as the first running boom. In the mid-1990s, the numbers leveled out before exploding again in a second running boom when the number of joggers rose spectacularly, not just in Europe and the United States but further afield as well. It was during this second boom that the commercialization of the sport really began to take off and new kinds of joggers started to join in, including women and middle-aged enthusiasts. Today, runners are a more diverse group than ever before.

The Netherlands boasts over 1.5 million recreational runners who run at least once a week, an amazingly high number for such a tiny country. In 2017, 56 million in the

United States said they ran regularly. That's nearly 17 percent of the total population. The percentages for the UK are comparable, with 10.4 million runners, or 16 percent of the population, in 2014. This includes people who run as little as four times a year, but the average British runner runs once or twice a week. In Australia, almost 22 percent of all adults surveyed in 2015 said they ran regularly, which translates to a total of over 5 million runners. So what is it about running that attracts so many willing participants? Sometimes a particular sport enjoys a temporary boom in popularity, as is currently the case with CrossFit and yoga. Running, however, seems to be much more than just cool or trendy; it has been gaining steadily in popularity for forty years and has reached the point where it is almost impossible to contemplate a world without joggers crisscrossing the landscape.

MOTIVATION

There is no shortage of information on what motivates runners to run. In 2008, the international market research company Synovate published the results of a comprehensive study of running in seven European countries. It revealed that people run for a variety of sometimes overlapping reasons. Fifty-four percent start running to stay fit, 40 percent to lose weight, and 21 percent to alleviate stress, while only 22 percent take it up just for fun. Most of these runners choose running above other sports because it is an outdoor activity that is easy to do, very flexible, and relatively inexpensive. The situation in the United States is quite similar, where 77 percent of runners say that staying fit and healthy is the most important reason why they run. They give other reasons, too, like losing weight and the decision to enter a particular race. And runners in the United States also run just for fun and to keep their stress levels down.

This all sounds very practical: running as a cheap and easy way of staying healthy. But does that fully explain its popularity? There must be more to it than that. After all, running wasn't suddenly invented fifty years ago. It may be immensely popular today, but people have been running since long before anyone figured out that it was a good way to stay healthy. When tracing the origins of running, most roads lead first to ancient Greece and then to the marathon at the first modern Olympic Games in 1896. Competitive running, as a form of physical exercise, has been around a lot longer than jogging. But does long-distance running actually have its roots in competition or is it older still? It may be a very ancient activity, one that has been around since the origins of humankind. We might even have a built-in primal instinct to run. To answer these and other questions we need to go much further back than ancient Greece. And not just a few thousand years, but millions of years back in time.

THE EVOLUTION OF MARATHON RUNNING

Daniel Lieberman studies fossils and bones and conducts experiments in an effort to understand why the human body is built the way it is and why all its parts work the way they do. A professor of human evolutionary biology at Harvard University, one of his favorite pursuits is studying the unusual shape of the human head. Compared to other mammals, we have a short, vertical neck; small teeth; a short, round tongue; and a small, snoutless face tucked away under the frontal lobes of the brain. Only humans have such weird heads. Most of us who are familiar with professor Lieberman know him from his research into the role that running played in human evolution. In 2004, he and his colleague Dennis Bramble from the University of Utah wrote an article on this topic that made the cover of the magazine *Nature*. According to Lieberman, humans are perfectly adapted to run long distances.

In their much-cited article, Lieberman and Bramble take a close look at the skeleton of *Homo erectus*, a species of humans that appeared around 1.9 million years ago. They compare the bones with those of chimpanzees and *Australopithecus*, an extinct human species that roamed the Earth between 4.4 and 2 million years ago. According to Lieberman and Bramble, *Homo erectus* was the first of our ancestors that was able to run on two legs, based on the anatomy of its skeleton. It was also the first species with the physical structures required to be able to run long distances. *Homo erectus* had strong ligaments at the back of its neck that stabilized the head and prevented it from jerking back and forth while running. It had larger buttock muscles than the chimpanzee and *Australopithecus*, which kept the torso firmly in place above the pelvis, and a long Achilles tendon and an arched foot that could absorb elastic energy each time it hit the ground and release it when the foot was lifted again. *Homo erectus* also had long legs relative to its bodyweight, short toes that made for energy-efficient running, and large muscles that cushioned the impact of the foot on the ground.

For primitive humans, a strong and specialized skeleton on its own was not enough to allow them to run long distances. The searing heat of the open African savanna and its never-ending grasslands presented *Homo erectus* with a formidable challenge: how does the body get rid of the heat generated by running for hours under a burning sun? Compared to apes, humans are much better at dealing with this problem, thanks to the presence of sweat glands all over the body and the absence of a thick coat of fur. Further proof that *Homo erectus* was an excellent runner can be found in the fact that modern humans are much better at running long distances than most other animals. With the exception of horses, hyenas, and dogs, not many animals are capable of completing a marathon at a moderate and steady pace. Our closest relatives—chimpanzees—and other primates are incapable of doing so, in any case.

PERSISTENCE HUNTING

So *Homo erectus* was probably good at running long distances. But its skeleton also reveals that it was ideal for walking for miles on end, which complicates matters slightly. The fact that our ancestor could walk and run presents two possible scenarios: that running long distances was an important phase in human development or that running may have been no more than a by-product of *Homo*'s improved walking skills. Bramble and Lieberman do not believe that running developed by accident, however. They argue that the human skeleton's specialized structures are not simply the result of walking. For example, to be able to walk you don't need an intricate "suspension system" in the foot that is capable of absorbing and releasing energy. And why did we develop such strong buttock muscles? When we run, these muscles work together to keep our torso steady, but they are far less active when we walk.

When you weigh up all the evidence, in Lieberman's opinion there can be no doubt that *Homo* evolved to be able to run long distances. He also believes he knows why. Primitive man started running so that he could source protein-rich food like meat, bone marrow, and brains. Not unlike a vulture, he ran around in the hope of finding carcasses before others carnivores discovered them first. Running may also have proved very useful during hunting. It is possible that *Homo erectus* was able to run after other mammals for so long that the animals, with their thick fur and lack of sweat glands, eventually collapsed with exhaustion from the heat. This is known as *persistence hunting*, a method whereby hunters who are slower than their prey over short distances are capable of wearing them down by keeping the hunt going for hours. Wild dogs, hyenas, and wolves are also good at persistence hunting.

Most experts agree that, thanks to their fat- and protein-rich diet, humans now have a unique kind of body with relatively short intestines, large brains, and small teeth. If it is true that the only way our ancestors could get their hands on meat was by covering long distances, then running is the reason why our bodies look like they do now. You could even say that running made us what we are—that it is literally part of our makeup.

It is a fascinating thought. Could running really be in our genes? Well, not everyone agrees with Lieberman's theory. Among anthropologists and evolutionary biologists there is widespread consensus that our large brains can be traced back to when the first humans started to eat lots of protein. However, whether *Homo erectus* started to run as a means of hunting is still a topic of much debate. The journalist and runner Christopher McDougall is a firm believer in Lieberman's untested hypothesis. In 2009, he published his international bestseller *Born to Run* in which he describes running as the superpower that turned humans into what we are today.

POOR SPRINTERS

When I visit Lieberman at Harvard University, I quickly discover that his ideas haven't changed much since the publication of his article in 2004. I meet him in his office in the Peabody Museum of Archaeology and Ethnology. The room looks like an extension of the museum and when I walk in I am greeted by a collection of monkey skulls grinning broadly at me from a display case. A collection of human foot bones lies scattered across Lieberman's desk. "I believe that primitive humans ran to put food on the table," he says when we are sitting across from each other in a corner of the room. "I think we were all born to run, in fact." But if humans have evolved to become long distance runners, why are some athletes very good at sprinting but not good at running marathons? Take Usain Bolt, for example: incredibly fast over short distances. Doesn't that mean he's a born sprinter? "We haven't lost the ability to sprint," Lieberman explains. "If you ever find yourself being chased by a lion, I'd expect you to run away as fast as you can. My point is that, compared to other animals, we are not very good at sprinting. However, you only need to be able to run faster than the person beside you to avoid being eaten by the lion."

This is true; compared to other animals, humans are pretty poor sprinters. Even the very best human sprinters can't run any faster than 10.2 meters per second, and then only for a maximum of fifteen seconds in one burst. By contrast, horses, greyhounds, and antelopes can gallop for minutes on end at speeds of up to 20 meters per second. For a long time, evolutionary scientists paid very little attention to running, primarily because of our mediocre sprinting ability. Until, that is, Lieberman and his colleague, Dennis Bramble, decided to study running from a very different perspective.

"The idea that our body is built with the primary aim of achieving maximum long-endurance capability is not true," he explains. "More speed means less stamina, and vice versa—a well-known trade-off in physiology. There is no doubt that natural selection has pushed us more in the direction of long-endurance capability, but that doesn't mean there is no variation. The best sprinters have a higher percentage of 'fast' muscle fibers, but they are not great marathon runners, and vice versa." Some athletes, however, are good at both kinds of running. Lieberman points to the Dutch football team as an example: "Those guys are extremely good endurance runners, but they can also run very fast when they need to. They are better at both long-distance running and sprinting than most of the rest of us. However, even they have to deal with the trade-off. The anatomy of your limbs and the ratio of 'fast' to 'slow' muscle fibers affects your ability to do one or the other."

If his theory is correct and humans are born to run, why do we suffer so many problems with our knees, shins, and hips? Were our ancestors somehow immune to injury when they were running around the savannas? "No, our ancestors also suffered injuries as a result of running. In fact, it's ridiculous to think they didn't!" Lieberman replies. Books like *Born to Run* have helped to create an image of the hunter-gatherer as the ideal kind of runner, one with almost supernatural running capabilities, much to the annoyance of Lieberman, who claims it is nothing but a myth. "Each time you use your body to move, you expose yourself to forces that increase the risk of injury. Of course, not moving at all leads to complaints, too, but of a completely different kind, like loss of muscle tissue, etc. I believe that humans ran a lot in the past and suffered injuries as a result, too. It's an inevitable part of life; every form of physical activity has its pros and cons."

The benefits attached to running were obviously greater than the consequences of suffering an injury. But do we suffer more injuries today than our ancestors did long ago? "That's an important question," says Lieberman. "If it is true that people suffer more injuries now, why is that the case? Is it because runners have become 'weekend warriors,' the kind who take no exercise during the rest of the week? Or is it because we spend most of our day sitting down and our hip muscles have become shorter as a result? Is it because of the shoes we wear? No one has come up with the answer yet."

PREHISTORIC RUNNING INJURIES

To find out more about prehistoric injuries, I arrange to meet the US paleoanthropologist Jeremy DeSilva. He studies how the first apes and our earliest ancestors propelled themselves, in particular how they used their feet and ankles. DeSilva argues that the explanation for many of the foot and leg injuries that humans suffer can be found in our evolutionary history. So is that also the case with running injuries? Do we suffer injuries because of some fault in our evolution? The Italian genius Leonardo da Vinci was a great admirer of the human body and he regarded the foot as the pinnacle of human evolution. In the fifteenth century, Leonardo dissected dozens of corpses in order to study the mechanics of the human body and the relationship between mechanics and function. He would take a foot and remove all of the flesh, tendons, muscles, and other soft tissue so that he could sketch the bones. Leonardo was always impressed by what he saw: "The human foot is a masterpiece of engineering and a work of art."

DeSilva holds a different view, however. "It all went wrong around four million years ago when our distant ancestors decided to start walking upright," he tells me. Fossils from that time reveal a whole range of prehistoric leg injuries, especially of the acute kind, such as sprained ankles and broken bones. Early humans were also probably prone to overuse injuries, but many of these kinds of injuries affect tendons and muscles instead of the bones. They are also more difficult to identify because soft tissue decomposes quickly. "However, we have been able to identify stress fractures in foot bones found in Spain that date back 800,000 years." And the reason behind all these injuries? Evolution. The human foot as we know it today evolved from the foot that apes used primarily for moving around in trees.

The transition from climbing in trees to walking upright on the ground has left its mark on the human foot. As tree dwellers, we had flexible feet with toes that could grip while climbing. That all changed when we moved down to the ground and our feet became stiffer so that they could support more weight. Our big toe retreated to line up with our other toes, creating a kind of lever that we now use to push off against the ground when walking or running. The first humans to walk upright also developed an arched foot, and the middle of the foot became stiffer, too. These changes facilitated improved propulsion—a flexible foot cannot push off as strongly from the ground. The heel also underwent changes; apes have a very small heel and the human heel is enormous by comparison. A large heel helps to distribute the impact on the foot evenly when it hits the ground. "We retained all twenty-six of the bones in our feet during this transition period," says DeSilva. "A thick layer of muscles and ligaments formed to help keep these bones together and stiffen up the foot. Evolution using plasters and paperclips, if you like."

The changes to the foot that enabled us to run on two legs were subtle enough, but the consequences were huge. Soft tissue like muscles and ligaments now have the job of absorbing the impact, which can cause the tissue to tear. The human foot also has to deal with the fact that our weight is now distributed over two limbs instead of four. In addition, we sprain our ankles very easily because they are so flexible—a direct result of our previous life up in the trees. Millions of years of tinkering has left us with an injury-prone and higgeldy-piggeldy human foot. "The problems of the past have not gone away," says DeSilva. "We are still walking around on a souped-up version of an ape's foot. Not exactly a designer's dream, in any event."

Nevertheless, life as a biped must have had its advantages for our distant ancestors. There has been much debate about why humans made the switch to walking upright. Charles

Darwin's argument was that it freed up the hands, while others claim that it made it easier for us to pick fruit and allowed us to look out over the tall grass of the savanna. Another theory is that it meant that a smaller portion of the body was exposed to the sun. Maybe it was just that walking on two legs was much more efficient and conserved energy. The only obvious drawback was the injuries caused to our legs and feet.

BLADE RUNNERS

So who, in DeSilva's opinion, are the number one bipeds? The answer: ostriches. Where we have twenty-six foot bones, the ostrich has only seven. "The bones of their foot and ankle are fused together to form one bone with toes on the end, which offers much more stability." Most large land birds don't have the kinds of foot and ankle muscles that can tear easily. Instead, they have strong ligaments that are excellent for storing elastic energy. "If they were allowed to enter the Olympics, ostriches would win every single race," according to DeSilva. It is no coincidence, therefore, that the prosthetics worn by blade runners are inspired by the legs of an ostrich. The prosthetics consist of a single piece of carbon fiber that bends and recoils with each step. The material is able to store a lot of elastic energy which propels the runner forward each time the blade recoils.

This is the job normally performed by our tendons. The Achilles tendon is able to store around 35 percent of the kinetic energy generated and use it to take the next step, while the arch under the foot saves around 17 percent. This spring mechanism allows a runner to recycle between

40 and 50 percent of the energy required to take their next step, although there is much discussion regarding the actual percentage.

The physiologist and biomechanics expert Peter Weyand and his colleagues at the Southern Methodist University in Texas were able to determine that the South African blade runner Oscar Pistorius used 17 percent less energy than the best able-bodied sprinters. A year before their findings, and at the request of the International Association of Athletics Federations (since 2019 known as World Athletics), Pistorius had taken part in a series of tests at the German Sport University in Cologne. The results showed that he reused three times as much energy as nondisabled sprinters. Pistorius lost only 9 percent of his energy each time he came in contact with the ground, whereas athletes with nondisabled legs registered a 40 percent loss of energy at the same constant speed. It appears, therefore, that prosthetics can provide athletes with a considerable mechanical advantage. However, the jury is still out on whether or not blades lead to an unfair advantage during competition. For instance, using the blades requires an athlete to draw twice as much power from their buttock and upper leg muscles as a nondisabled sprinter does. The most efficient mechanical alternative bears little or no resemblance to the anatomy of the modern human foot, in any case. "Our feet are good enough to do the job they are supposed to do; otherwise humans would have vanished from the face of the earth a long time ago," says DeSilva. In other words, they basically helped us to survive. Comfort, however, is a different matter altogether.

Today there are huge variations in the way people walk and run. DeSilva believes that in prehistoric times there were fewer cases of flatfoot and turned-in feet. Back then, the shape and function of your feet probably had a significant bearing on your chances of survival, but such factors are of little consequence nowadays: "Today, it is unlikely that a person who runs less efficiently than others will end up being eaten by a lion. That simply doesn't happen anymore. With the result that the differences between how people walk and run are bigger than ever before." At least, that is, in Western society, where we are used to wearing shoes. "These variations are less prevalent in cultures where people still walk around on their bare feet."

So, despite our susceptibility to injury, is it still a good idea for us to run long distances? DeSilva doesn't see any problem with it, because we have a safety net: "Walking on two legs was a tricky business in the past sometimes. This manner of self-propulsion could only have evolved in a social kind of animal." Fossils have been found of sprained ankles that obviously had time to heal again. How could that have been the case back in prehistoric times, with lions lurking around every corner? If you sprain your ankle today, you can be out of action for weeks. "The only logical explanation is that our ancestors took care of each other." Something we still do today, thankfully. An injury can be very inconvenient, but it doesn't have to be a death sentence.

BACK IN FASHION

Whether being able to run long distances was an essential aspect of our evolution or not, the fact is that we are good at it. For proof we need look no further than the Tarahumara in northwestern Mexico. Anthropologists have long been fascinated by this reclusive tribe and their athletic prowess. Christopher McDougall has written in great detail about this tribe of runners after spending some time with them in Copper Canyon in the Mexican state of Chihuahua. If you are looking for the best runners on the planet, the Tarahumara are your safest bet. They run long distances for the purposes of communication, transport, and hunting, often covering up to two hundred miles in less than two days. Sometimes there is necessity attached to the reason for running, for example in order to deliver a message. But there is often an element of play to it, too, and both men and women from the tribe can spend hours or even days running after a wooden ball.

The Tarahumara normally hunt with a bow and arrow, but anthropologists have also witnessed them hunting deer or turkeys simply by running after them for a day or two until the animal collapses in exhaustion, after which they strangle them with their bare hands. So humans practice persistence hunting, too; the Tarahumara do, in any case. Tribes in the Kalahari Desert are also known to participate in persistence hunting, but it is an extremely rare form among modern hunter-gatherers, according to the US anthropologists Travis Pickering and Henry Bunn in an article published in the *Journal of Human Evolution*.

Our current fascination with running may be a by-product of two million years of human evolution, but there was a lengthy period in which we hardly did any long-distance running at all. When agriculture began to establish a firm foothold in society around ten thousand years ago, our ancestors no longer needed to roam around all day looking for food. In fact, you would have been considered crazy if you went for a 26-mile run after spending the whole day plowing your field. Despite the emergence of agriculture, however, it appears that we retained our ability to run long distances. In any event, running is now well and truly back in fashion, albeit in a different guise. It certainly has nothing to do anymore with finding protein-rich food. In fact, the opposite is now the case: today, runners consume proteins with the aim of keeping their muscles strong enough for the many miles still ahead.

FUN FACTOR

Some researchers believe that we are now in the middle of a third running boom, one characterized by the growing number of so-called fun runs, such as 5K races combined with a fun factor. One example is the Electric Run in which the runners and the course are illuminated by ultraviolet light. Another popular event is the Color Run, which was first organized in 2011 in Phoenix, Arizona. At each kilometer post, the runners are doused in a different color powder, a practice borrowed from an old Indian tradition. There are also runs that are combined with other sports, such as obstacle runs in which you have to climb, jump, and crawl as well. These events are very popular with young people. In fun runs, it is more important to enjoy yourself than to run a fast time; sometimes the run isn't even timed at all. I have taken part in a survival run on two occasions. In one five-miler, we had to tackle forty-five obstacles that required far too much rope-climbing for my liking. I spent most of my time either up to my knees in mud or trying to swing across a ditch on a long rope. It was great fun—no doubt about it—but give me an old-fashioned run in the park any day.

In my preliminary research I was able to establish that, even if we have evolved to be able to run long distances, our evolutionary history means we are condemned to suffer injuries as a result, whether we like it or not. And that's a bit of a bummer. We runners will just have to make do with the tools we've got and accept that suffering injuries is simply part of our makeup. But that doesn't mean there's nothing we can do to limit the risk of injury.

My search for the science behind running begins in earnest with an examination of the body that evolution has handed down to us. Are there any tricks or tools we can use to get that body to function more efficiently while running? The first things that spring to mind are running shoes, nutrition, and mental training. And in what ways do our limbs, organs, and genes affect our capabilities? I'll start by addressing the one thing that nobody seems to be able to agree on: good running shoes.

2

THE USEFULNESS OF RUNNING SHOES

Want a strong, solid relationship that is willing to go the distance? Get to know your running shoes.

—Dean Karnazes, ultramarathon runner

I wore tennis shoes for the first few months of my running career. But as soon as I knew I was hooked, I went to the nearest sports shop and bought the first pair of running shoes that caught my eye. And that was that. It took another two years before I decided to invest more time and money in finding the right kind of shoes in a proper running shoes store. I had just suffered my first knee injury, so it seemed like the wise thing to do. At a reputable outlet, the sales assistant brought me a few different pairs to try on. He then got me to run in a straight line down the length of the store and filmed me with a camera installed on the floor. We watched the resulting clip together. "Your ankles tend to turn in when your foot hits the floor," he said. Not good news, apparently. Fortunately, my ankles behaved much better in a pair of running shoes that offered more support; they still turned in, but to a lesser extent. I still remember how self-confident I felt walking out the door carrying my first-ever pair of customized running shoes, as if I had just made the switch from beginner to advanced runner. Shortly afterward, however, I was struck down by my first serious knee injury. And yet I continue to buy the same ASICS shoes to shore up my feeble ankles.

The idea is that good quality running shoes, preferably tailor-made to fit your feet, can help prevent injury. But do they really make a difference? The extra support in the sole has never helped me with my dodgy knees or weak hamstrings, let alone the awful shin splints I suffered for six months. Trainers and fellow runners often give me well-intentioned advice to have my shoes checked out. It is almost an accepted fact in the world of running: if you are prone to injury, it's probably because of your footwear. However, the fact that everyone accepts a certain assumption doesn't make it true. But can running shoes actually protect us from injury? That's the million-dollar question. Let's see what science has to say about it.

INJURY PRONE

It was during the running boom in the 1970s that the first articles concerning injuries began to appear in scientific journals. Studies carried out over the past fifty years have thrown up very diverse sets of statistics. Some show that 20 percent of all runners suffer injury on an annual basis, while for others the figure is as high as 75 percent. So why is there such a large discrepancy between the figures? We can find the answer in a review cowritten by biomechanical

engineer Benno Maurus Nigg at the University of Calgary in
Canada. First of all, runners today are quite different from
the runners of the seventies and eighties. The latter were
dedicated runners—they wanted to be winners, were invari-
ably of slight build, and running was their primary sport.
In addition, three-quarters of them were men. The picture
is much different after 2000 when the majority is made up
of runners who run for recreational purposes and who take
part in races like marathons with the aim of finishing rather
than winning. Some are overweight and most are active in
other sports too. And, significantly, a slight majority of all
runners today are women. It must be pointed out, however,
that some studies base their results on beginners while others focus on competitive runners. Studying different groups
on a selective basis cannot provide us with a reliable picture
of changes in the frequency of injury among runners.

Another reason why research is often hampered is that
everyone has a different definition of the term "injury." For
some scientists, an injury is something that causes the runner to seek specific medical help. Others refer to every single
symptom of pain and discomfort as an injury, and these
kinds of complaints are far more common than injuries
requiring medical treatment. In any case, we simply cannot
compare the figures from the 1970s with those of today. As
a result, we cannot say much about whether the number of
injuries has risen or fallen over the past fifty years, apart from
the fact, of course, that runners are a breed apart. They are
renowned for being highly prone to injury. In 2018, running
had the second highest number of injuries after soccer in the
Netherlands: six cases of injury for every one thousand running hours, almost two times more than the average number

of injuries for all other sports combined (excluding soccer). I am beginning to wonder whether good quality running shoes offer us any protection at all.

AIR BUBBLES

There was a time when only trained athletes in excellent physical condition ran marathons. And they didn't do so wearing multicolored shoes with soles like bouncy castles, either. Their feet were usually supported by nothing more than a few strips of rubber and cotton sown together. So why did this simple model suddenly disappear? In the 1970s, the number of recreational runners began to explode, which led to the meteoric rise of the running shoe industry. People had started jogging just for fun, and they wanted to be able to do so in comfort.

You can see the results of fifty years of innovation for yourself by walking into any good running shoes store—wall-to-wall shoes in all kinds of colors and models. The names alone are evidence of a rich imagination: the "Adrenaline GTS," the "Gel Quantum," and the "Downshifter." One of the best monikers ever has to be "Gel Hyper Speed," as if these shoes offer you the ability to run at the speed of light. It's wonderful to have such a huge range to choose from, but the assortment overdose has given rise to one of the most crucial questions in the world of running: which shoes are the best?

The first major technical innovations were introduced back in 1979 with the launch of the Nike Air, the first shoe with air bubbles in the sole. The inventors added the bubbles because they figured that running on hard surfaces meant that the legs were subjected to a high level of impact. They also believed that repetitive impact was a significant cause of injury. Ever since, the idea that cushioned soles are indispensable has been drilled into the mind of every runner on the planet.

Runners do have to absorb a lot of impact when running; the shockwaves travel upward from the feet to the rest of the body until they reach the runner's head, where they can't go any further. It may not be immediately apparent to the average runner, but tests on treadmills equipped with a built-in force plate—a system used to analyze running technique—show just how big those shockwaves can be: with each step, and depending on your speed, your leg is exposed to a force up to three times greater than your own bodyweight. And when you increase the tempo, the impact is greater than when you are jogging easily. Generally speaking, the impact when running is twice as hard as what you experience when you are walking. When you are running, you take around 160 steps per minute. That adds up to 4,800 steps in half an hour. And remember, each step sends the equivalent of three times your bodyweight shooting up through your leg. For a thirty-minute run that means a lot of impact. Wouldn't it be nice if at least some of that impact was absorbed by something other than your body? Air bubbles in the sole seemed to offer the promise of relief. Ever since the Nike Air appeared, shoe manufacturers have been competing fiercely with each other to develop the ultimate shock-absorbing material, and shoes that offer a cushioning effect still dominate the market today.

FEEBLE ANKLES

A second technology that has become part and parcel of the world of running shoes stems from the 1990s, when the shoe manufacturer Saucony introduced its dual density sole technology. In running shoes made using this technology, the midsole is harder under the arch of the foot and offers more support to the inside of the foot, which prevents it from turning in. In stores these are known as motion control shoes.

Feet and ankles that tend to turn in are bad news for your knees. According to data from Running Warehouse, an online store for running enthusiasts, the feet of 50 to 60 percent of all runners tend to turn in too much. Everyone has this tendency, which is known as *pronation*, to some degree or other, and that's a good thing because it helps to spread the shock of the impact when the foot hits the ground. My ankles tend to turn in too much, and that makes me an *overpronator*. Some 20 to 30 percent of all overpronators pronate to an excessive degree. A quarter of all runners run the way they ought to and are therefore "neutral." At the other end of the scale are runners whose feet turn out instead of in, a less common phenomenon known as *supination*.

It has been suggested more than once by scientists that overpronation is closely linked to the risk of injury, hence the practice in stores of filming a person's gait in an attempt to find the shoe that best fits the type of foot. However, there is also evidence for the opposite: that there is no relationship between overpronation and injuries. "Our capacity to prevent such injuries is currently limited, with training advice and footwear prescription forming the mainstays," wrote exercise physiologist Robin Callister in the *British Journal of Sports Medicine*. A neutral foot position is the best position, and so runners like me would be well advised to wear motion control shoes when running. In theory, at least. I have to admit that the extra support and cushioned sole do give me a secure feeling. But this still does not answer the question of whether they can prevent injury.

COMPARING SHOES

Today I am heading off to put the theory to the test. I have jumped on a train armed with a bag of running shoes and am on my way to the movement laboratory at the Catholic University of Leuven in Belgium. The lab is located right next to the running track at the Heverlee sports center. I have an appointment with Ellen Maas, a PhD student from the biomechanics research team on the Faculty of Human Movement Sciences. Together with her colleague Kurt Schütte, she is going to prepare me for a comprehensive gait analysis. My aim is to find out what kind of shoes are best for me.

After Maas and Schütte have finished prepping me, I resemble a patient who has just had surgery to remove varicose veins. White nets have been attached at strategic points to my calves, knees, and thighs. Their job is to hold the small reflective balls in place that the scientists will be sticking onto my body and shoes. Ten infrared cameras positioned all around the treadmill will soon be tracking the movements of those balls while I run. The analysis in this lab goes a lot further than the video clip that was made of my feet in the shoe store. Everything will be monitored and measured: the forces exerted on my body and the angle of my limbs while running. And this is done not once but twice—first while I'm wearing motion control shoes and then wearing neutral shoes.

"Your kneecaps are turned in," says research team leader Benedicte Vanwanseele from the Faculty of Movement and Rehabilitation Sciences, who has popped into the lab to see what's going on. "Do you ever have problems with your knees?" That's a great start. I've barely stepped onto the treadmill and my chances of walking out of here with a good gait are already being written off. Vanwanseele believes that shoes have the potential to both cause *and* prevent injuries. But whether they help or not depends on the individual. "Not everyone reacts the same way to the same shoe," she explains. Her studies have shown that while a pair of motion control shoes might help one person's gait, they may have no effect whatsoever on the next. The burning question then is: What kind of shoe works for which kind of person? Can you establish that simply by making a video in a specialist shoe store? Vanwanseele immediately looks skeptical. "In a store, all they look at is pronation, but that is only one aspect of how a person walks." I know what she means. The minute your ankles show even the slightest tendency to turn in, bingo! Another pair of motion control shoes sold. "A short video like that doesn't tell you anything about how the foot behaves inside the shoe either," adds Vanwanseele. "You can't see through the material, after all." Experiments have shown that the ankle also moves inside a running shoe. In other words, on the video it might look like there is less pronation thanks to the motion control shoes, but in reality the situation is often more complicated.

EXPENSIVE MEANS WORSE?

In 2015, Vanwanseele was invited to appear on the Dutch TV show *Radar* to talk about running shoes. She carried out a live test of cheap and expensive sports shoes and even took the soles apart to see what they were made of. The expensive Nike shoe had more layers of rubber in the sole. Surprisingly, however, the shoe from the discount Bristol store, with its simple sole, actually provided better cushioning. Ellen Maas, who also appeared on the show, was not happy with the way the subject was handled, she explains, as she tapes the reflective balls to my body. "*Radar* wanted to see conclusive results within a few hours, and of course that's not possible. It usually takes a few days to carry out that kind of analysis." So the conclusion arrived at on TV that the cheap shoes were better than the expensive ones has to be taken with a pinch of salt.

The idea that expensive sports shoes are actually worse than cheaper ones has been making the rounds for some time now. Research carried out by Bernard Marti and his colleagues at the University of Bern is often cited as proof of this claim. In 1984, Marti and his team asked 4,358 participants in a race to fill out a comprehensive questionnaire. Each runner provided information about their training routines and footwear over the past year. Almost half of them indicated that they had suffered an injury in the previous twelve months. And, believe it or not, there seemed to be a link between expensive shoes ($140 or more) and susceptibility to injury. But what was the cause and what was the consequence in these cases? It proved impossible to say. It may simply be that runners who are more injury-prone, or those who eat up lots of miles, invest more often in expensive shoes, according to Marti. He concluded that it was more or less impossible to establish any correlation between the price of a pair of shoes and the risk of injury.

For the test in Leuven, I have brought along an expensive and a cheap pair of running shoes, but I am primarily interested in finding out the difference between them in terms of model. A few weeks ago, without any advice from an assistant, I bought a pair of Nikes with no motion control features at a discount store. Buying the shoes was a whole new experience in itself, as it took forever to find someone who could fetch the second shoe for me from the back of the store. My newly acquired Nikes do have a fairly thick cushioned sole, and we are going to compare them with the pair of ASICS motion control shoes that I purchased on the advice of the assistant at the specialist store (for nearly three times the price).

The analysis is carried out as follows: first, five minutes walking to get used to the treadmill, then two minutes jogging at 7 km/h, followed by two minutes running at 8 km/h. "The pace might seem slow to you, but running on a treadmill feels faster than running outdoors," says Maas. The treadmill has a built-in force plate and is situated in the middle of the lab. With each step I take, the force of the impact on my body is measured and sent to the computer. Schütte has also attached an accelerometer to the insides of my ankles and to my lower back. These will gauge how quickly the shockwaves move up my leg and when and where they begin to diminish. He tells me that the impact of running is absorbed by shoes, tendons, and muscles in order to protect the brain.

Finally, it's all systems go. "Ready for the first test?" asks Maas. I nod, jump into my ASICS, and step up onto the treadmill, which then starts with a jolt. I walk, jog, and then jog a little faster. So far so good. The balls attached to my body reflect the light back to the cameras, with the interval depending on their relative distance from the cameras. The computer screen is projected on the wall so that I can follow the movements of the balls in real time. It all looks pretty good to me. After another few minutes of jogging, the first test is done. The treadmill comes to an abrupt stop and I almost lose my balance. I change into my other shoes and the balls are transferred from my ASICS to the Nikes. The treadmill starts rolling again. These shoes feel comfortable too. I complete the second test without any trouble, the only difference being that this time I am prepared for the abrupt stop at the end. I can expect the results in two weeks' time.

FIT FOR THE TRASH

Someone who thinks very differently from Vanwanseele when it comes to the usefulness of running shoes is Steef Bredeweg. He is a sports physician and director of the University Sports Medical Center in Groningen in the Netherlands. He has had injured runners coming to him for years, and in 2014 he did his doctoral thesis on the cause and prevention of running injuries. His conclusion? To date, we have no evidence to suggest that risk factors like age, gender, BMI, background, shoe type, or foot type play a major role. The only factors that appear to significantly increase the risk of injury are the volume of training and previous injuries.

Although his research was not focused exclusively on running shoes, he is very familiar with the relevant literature. At conferences at which he is invited to speak, he is not afraid to suggest that running shoes are of no help at all and are really only fit for the trash. I am sitting opposite him at the desk in his office. Bredeweg tells me that he was also invited to appear on the TV show *Radar* to talk about running shoes, but he declined. "Journalists, especially those on TV, only want to hear you say that it's all bullshit," he says. But isn't that precisely what he claims himself? Running analyses, all that fuss about data, choosing the right shoes—he certainly has his doubts. He tells me that ASICS also has a treadmill in Amsterdam where customers can take a biomechanical test. "People like to be shown the numbers that tell them they are doing the right thing, I get that. A running test in itself is no harm, of course, but the results are far from scientific. The manufacturers' primary motivation is to hoodwink their customers into buying more shoes."

Shoes designed especially for women, shock absorption, motion control, a sole with an elevated heel—in his opinion they really don't make much difference. What about the advice that you should buy a new pair of shoes after every five hundred miles because the spring will be literally gone out of your step? Also nonsense, says Bredeweg. Neither is he convinced of the necessity of a cushioned sole. Of course, there is certainly a link between heavy impact and stress fractures in the shinbones, he adds. But that is the only obvious link. The evidence for other injuries is far from convincing.

THE SHOE ENIGMA

Many researchers believe that the shocks that our legs repeatedly absorb are one of the main causes of injury. From a biomechanical point of view, this seems entirely logical. If you have ever struck your finger with a hammer, then you know how impact can lead to injury. So taking thousands of steps one after another must surely result in some kind of damage to our feet; after all, they have to take the brunt of the blow. Surprisingly, however, science has very little data on the relationship between impact and injury. The only obvious link is the one between impact and stress fractures to the shins and toes mentioned above: when bones are exposed to strong forces, they can develop tiny cracks, just like wood splits when you drive a nail into it. But the most common injuries among runners are knee injuries. Furthermore, the results of comprehensive studies do not point to the impact of landing as a major risk factor with regard to injury.

According to Benno Maurus Nigg, in the 1970s scientists just assumed—without any epidemiological evidence whatsoever—that the physical impact of running caused injury. In 2014, Nigg and his colleagues dispelled a number of myths surrounding running shoes in an article in the *British Journal of Sports Medicine*. The fact is that, up to now, research has not been able to prove conclusively the relationship between impact and injury. If increased impact is in fact related to the risk of injury, then in theory you would expect the fastest runners to suffer the most, according to Nigg. Which sounds logical too: the faster you run, the greater the impact with the ground. But we don't even have any anecdotal evidence to suggest that the faster you run the more likely you are to get injured, let alone evidence based on scientific observation.

This is not the only enigma surrounding the running shoe. For instance, the extent to which shoes with cushioned soles actually absorb the impact of running is still very unclear. A sole with air bubbles usually softens the blow a little, but is it enough to make any real difference? "You can put an egg inside an oven mitt, but it will still break if you hit it with a hammer," as Christopher McDougall explains in *Born to Run*. Many runners buy new shoes every year in the belief that the cushioning effect has disappeared from their old ones—and this assumption is correct. The sole becomes stiffer with age, and there is evidence that the stiffer the sole the greater the impact on your legs. Even if you don't use your shoes for a while, the material will continue to age.

There might be a drop in the cushioning effect, but according to Bredeweg that is no reason to go running to the nearest sports shop for a new pair of shoes. "A good reason for buying new shoes is when your old ones aren't comfortable anymore. Or when the inside is beginning to show a lot of wear and tear," he says nonchalantly. Is it really that simple? The more time you spend with Bredeweg, the more you believe that the usefulness of running shoes is extremely overrated. Of course, shoes offer great protection against sharp stones and glass. But there is very little difference between one pair and the next, and so we can dispense with all those commercial tests and videos in the shoe store. So what should we look for then when buying running shoes? Bredeweg gives me a one-word answer: "Comfort," before adding: "As long as they are comfortable, they'll do. That's the best we can expect, based on current knowledge."

In his doctoral thesis, Bredeweg suggests that matching certain kinds of shoes with certain kinds of runners results primarily in a false sense of security, thereby actually increasing the risk of injury. "It is only a theory," he emphasizes. "If you buy a large and very safe car, your driving behavior is going to change accordingly." Or take the skin cancer paradox: people who use sunscreen have a higher risk of cancer, simply because they feel safer and end up spending more time in the sun. The same principle applies to shoes. "The feeling that the shoes you are wearing were made to fit your feet makes it tempting to believe that you can run any way you like in them."

LACK OF EVIDENCE

"Is Your Prescription of Distance Running Shoes Evidence-Based?" This is the title of a study published in 2008 in the *British Journal of Sports Medicine* by exercise physiologist Robin Callister at Newcastle University. Her colleague, Dr. Craig Richards, was also involved in the study. Richards, a runner himself, had become completely fed up with getting injured all the time despite the fact that there was nothing wrong with his own biomechanical setup. And he wore "good" shoes too—didn't he? Richards began to blame his chronic complaints on his footwear, and so one day he decided to try something different. He left his shoes at home and taught himself how to run barefoot. And hey presto! His injuries vanished into thin air. This little miracle prompted him to take a closer look at the modern running shoe.

The researchers wanted to find out whether the current practice of prescribing shoes on the basis of a person's foot type was based on solid fact. They consulted the literature and all of the studies they could find on modern running shoes and injuries and were shocked at what they discovered. No one had ever conducted a clinical study (i.e., a randomized study using a control group) into the effects of modern running shoes on the risk of injury and performance. And yet almost every runner believes that they need stuff like shock absorption and motion control. Even professionals like physiotherapists are convinced of this. Callister and Richards were left with little choice but to conclude that in evidence-based practice we have no grounds for advising runners to wear motion control shoes with an elevated, shock-absorbing heel. Ironically enough, I bought my first-ever pair of anti-pronation running shoes at almost the exact same time as their article was published.

Despite the lack of evidence, up until recently fancy hi-tech running shoes were included as important accessories in the official guidelines of the American College of Sports Medicine (ACSM), the largest sports science organization in the world. This means that every physiotherapist and sports physician who had complied with these guidelines had been trying to match shoes to foot type without any sound scientific reason for doing so. Other influential organizations, like Sports Medicine Australia, have openly admitted to tailoring their advice on running shoes to fit the terms of their sponsorship deals with ASICS. ACSM has since revised its recommendations with regard to shoes and added a little more science to the recipe.

It took until 2012 for the first substantial clinical study to be carried out into shoes and injuries, and the preliminary results have since been published. A group of Danish scientists conducted a study of one thousand runners who all started out wearing neutral shoes. After one year, the over-pronators reported the least number of injuries, although you would expect the opposite to be the case because they were the runners who needed extra correction. However, the difference wasn't so large as to be able to rule out chance completely. Nevertheless, it appears that overpronators like me don't necessarily need to wear motion control shoes. This all seems to point to an obvious conclusion: modern running shoes do not reduce the risk of injury.

INVINCIBLE DOGMA

In the meantime, I am itching to find out the results of my running test. Up until now, research has not been able to provide much evidence for the idea that some shoes are more beneficial to runners than others. Theoretically, we may not be able to say much yet about the benefits of anti-pronation technology, but we also need to look at what's happening out there on the ground. Two weeks after my trip to Leuven, Maas mails me a set of charts and videos. We go through the data together on Skype and quickly discover something unusual. We had both expected that my ankles would move around a lot less in the shoes with motion control. After all, they are designed to keep your feet firmly in place. The charts told a different story: we could not see any difference between the two pairs of shoes. In both the neutral and anti-pronation shoes, my ankle did the same thing: turn slightly inward. I had begun to have doubts about the technology after my conversation with Bredeweg, but this was still a big surprise. My "corrective" shoes didn't seem to be doing any correcting at all. Maas did add a note of caution, however. The sensors recording the movement were attached to my shoe and not my ankle, and so the cameras were only able to monitor how the shoe itself turned in and not the small movements that my foot made inside the shoe.

Nevertheless, whether the ankle follows the movement of the shoe precisely or not, or exhibits only slightly different movements, the last thing you would expect a *motion control* shoe to do is turn in.

We scroll down through the rest of the data from the analysis. Maas stops at the chart showing the impact of the treadmill on my legs. It appears that they are subjected to more impact when I am wearing my ASICS motion control shoes. I don't get it. How could that be? "It may be because the arch of your foot fails to fully absorb the shock," Maas explains. "The stiff part of the sole in your motion control shoes provides the arch with a lot more support, and this prevents it from relaxing and distributing the impact of the landing evenly when your foot hits the ground. In that case, the shock absorption has to be done mostly by the sole, but that doesn't seem to happen here." However, comparing the shock absorption capabilities of these two pairs of shoes is not exactly fair: my ASICS are a year older than my neutral Nikes and I have covered a lot more miles in them, so I can't expect the soles to provide as much cushioning as they used to.

We do identify a number of differences between the shoes, but nothing startling. The scientists in Leuven reach the conclusion that while there are slight differences between how I run in my neutral shoes and how I run in my motion control shoes, my gait looks perfectly normal in both cases. However, they cannot say much about the usefulness of cushioning and anti-pronation. Maas: "We simply don't know enough about the relationship between the technologies and injuries." Which means I have no reason to doubt the good intentions of the sales assistant at the store when he convinced me to buy my motion control shoes. After all, the idea that poor footwear can cause injury is a persistent one in the world of running. In fact, it is almost an invincible dogma.

THEORY AND PRACTICE

The story of the development of the running shoe is a rather odd one. Even before the first running boom, researchers had identified pronation as an important factor with regard to the risk of injury. The same applies to impact loading. But where was the evidence? However plausible these theories might sound, no scientific proof has ever been forthcoming. For the moment, all we can rely on is our practical experience on the ground. Anecdotal evidence does seem to confirm the usefulness of running shoes. Some runners claim that they experience less knee pain, for example, when they start using a new pair of shoes. However, it is impossible to say with any certainty that it is the shoes that are responsible for the improvement. When you suffer an injury you tend to change several things, like your pace and the distances you run; you become a bit more careful, in other words. And your positive expectations with regard to your new running shoes naturally play an important role, too.

It is never easy closing the gap between theory and practice. Scientists have studied large groups of people only to find out that the type of running shoe has no obvious effect on injuries. However, these studies cannot establish what a particular kind of shoe can do for an individual runner. Some probably benefit from certain models, as Benedicte Vanwanseele's research suggests. But that doesn't mean that the credibility of videos made in shoe stores—no matter how professional they might seem—does not depend on a semblance of scientific justification that simply does not exist.

I currently have three pairs of running shoes in my shoe rack at home. I will hold onto the two pairs that I tested in Leuven, even though there seems to be no difference between the "customized" pair and the pair I bought at the discount store. I bought the third and most recent pair without testing them first using a video clip. Like the first expensive pair, they are very colorful and futuristic-looking, but the sole isn't as thick. They are also less robust and a lot lighter. At first I was very disappointed when I found out that running shoes are not as useful as I had believed, but now I realize that this actually makes life a lot easier. When I go to buy my next pair, all I will have to do is check whether they are comfortable or not.

At the end of the day, my search for the best kind of running shoes did not deliver much in the way of concrete results. From a scientific point of view, the jury is still out. But what can science tell us about running without shoes? Some people believe that modern footwear actually does more harm than good; that not only do cushioned soles and motion control technology do nothing to prevent injuries, they can actually cause them instead; and that we would be better off running barefoot anyway, like our ancestors did. But is there any evidence to support the claim that barefoot running is better?

TIP

A tip from Steef Bredeweg (not scientifically proven, but based on his experiences with injured runners):

If you want to run regularly, for example, six miles two or three times a week, then variety is very important. The more varied your runs, the less likely you are to overburden specific parts of your body. Use different routes and run on different surfaces, and rotate your shoes as well. If you are running between twenty and forty miles a week, you should have at least three different pairs of running shoes. When the first pair are worn out you can stop using them. You will still have two pairs of shoes that you are used to running in and you can start breaking in a new pair too.

3

BACK TO BARE FEET

How one runs probably is more important than what is on one's feet, but what is on one's feet may affect how one runs.
—Daniel E. Lieberman, Harvard University

If you pay enough attention, you are bound to spot them: runners jogging along with nothing on their feet, especially at popular running events. What makes someone dump their shoes and start running barefoot? While the trend may not be equally prevalent everywhere, barefoot running has established itself firmly as a subculture among runners. And that's not all. The media and sports science have also been paying close attention to the phenomenon. The debate surrounding barefoot running is one of extremes: you are either a true believer or you think it's all a load of baloney. Bloggers from both sides of the argument battle it out regularly online. Proponents of barefoot running (also known as natural running) claim that it is, well, more natural and therefore better for you. They argue that running in shoes not only doesn't help prevent injury, it can actually be harmful.

And that modern footwear can damage our knees, shins, and the soles of our feet. Advocates of "natural" running often defend it with a kind of evangelistic zeal, in the opinion of Ross Tucker, an exercise physiologist and running coach at the University of Cape Town.

According to opponents of the idea, however, barefoot running is both dangerous and unhealthy. They often use the argument that feet need protection and support in the modern environment of asphalt and concrete. These equally aggressive "anti-barefooters" dismiss those on the other side of the fence as fanatics. Both camps continue to fire anecdotes and scientific studies at each other in their efforts to prove each other wrong; no wonder barefoot running is such a bone of contention. But what can we say about it in scientific terms? And who's right? The believers or the skeptics?

EXEMPLARY ANCESTORS

When you read the blogs and stories that are posted online, two arguments in particular in favor of barefoot running emerge. The first is that it is inherently good because we used to do it in the past and it is more natural. The second argument is that shoes are bad for our feet. The word 'natural' doesn't really help here. Not everything that is natural is necessarily good for us. And not everything that is unnatural is bad. In fact, it is often the other way around. Caesarian sections, antidotes for snakebites, painkillers, and pacemakers are all very unnatural. Even the apples and broccoli that organic farmers grow are not entirely "natural" anymore. On the other hand, the most poisonous substances on earth are produced by Mother Nature herself. So is something automatically good for us just because it reflects the habits of our prehistoric ancestors? The evolutionary biologist Daniel Lieberman from Harvard University and his colleague Dennis Bramble argue that humans have been running long distances for over two million years, and for most of that time without the help of sports shoes. If their theory that the ability to run long distances was crucial to survival is correct, then it follows that the human body has adapted to run barefoot.

From an evolutionary perspective, wearing shoes while running could be classified as abnormal behavior. Ever since humans started to walk upright, they have done so 99.9 percent of the time without anything on their feet. You could regard this as reason enough to revert to our former mode of running. But "because that's the way our ancestors did it" is not a valid argument in itself. What was once good for prehistoric humans is not necessarily so in the modern age, as Lieberman explains in his book *The Story of the Human Body*. He uses the so-called paleo diet as an example: "The fact that we have evolved to eat certain kinds of food does not automatically mean that those foodstuffs are good for us or that there are no other kinds of food that might be better." Primitive humans were not averse to eating the contents of the stomach of an animal carcass if that helped them to survive. Fortunately, today our food is usually a lot more appetizing.

Prehistoric customs are not, by definition, the "way to go." We and the environment we live in have both changed dramatically since humans first began to run. Whereas our ancestors ran long distances over rough terrain to hunt animals for food, today we run mostly for fun and often on the

sidewalk or a running track. We cannot compare the circumstances then with those now, so what worked in the past
is not necessarily the best option in the present. In fact, it
may even be that humans are no longer really suited to running because of our modern lifestyle. Or maybe running was
never a good idea in the first place; after all, fossils show that
primitive humans also suffered running injuries.

The second argument—that modern footwear can actually
cause injuries—is a more interesting topic for the modern-
day runner. Are our bodies somehow not suited to wearing
shoes? Shoes with thick, cushioned soles whose purpose is to
keep the foot under control are designed to prevent injury
and to be a better alternative to running barefoot. But, the
argument goes, they do not help to prevent injury and they
may even do the exact opposite. So could the remedy be
actually worse than the ailment? Benedicte Vanwanseele
believes it could. In her opinion, motion control shoes that
support the arch of the foot are not a good idea if you have
no problems with the arch in the first place. "The danger is
that when you start fixing stuff in the foot that is not broken, the problem will only move further up the body," she
says. "Your knees may be adversely affected, for instance." It
may be that our feet are not so happy with all that superflu-
ous comfort after all.

THE SHOE AS CULPRIT

Physiotherapist and biomechanical engineer Irene Davis
at the Harvard Medical School doesn't care much for modern running shoes. She has spent the last twenty-five years
studying the link between biomechanics and injuries. In

2012, she set up the Spaulding National Running Center in Cambridge, Massachusetts. Today, runners from all over the United States come to her for help with their injuries. Since opening the center, she has treated more than five hundred patients and has studied hundreds of runners in her research. Davis believes that modern footwear weakens the muscles of our feet. Instead of training those muscles and keeping them strong, we turn to motion control shoes for help. A big mistake in her opinion.

On a cold December day, I meet up with Davis to discuss her research. Her studies of the human foot have been inspired by the patients who continue to stumble through her door in search of answers. "Every injured runner who comes in here has weak feet, and I believe that this can be attributed to the constant support that shoes provide," she begins. She is referring primarily to anti-pronation shoes, whose job is to offer extra support to the foot. The shoes feel very comfortable, thanks in part to the soft cushion in the sole, but maybe they are actually *too* comfortable. All that extra support may be preventing the foot from functioning the way it was designed to. "In all probability, shoes alter the way in which the foot muscles are utilized," she explains. "This is the reason why people who always wear shoes cannot spread their toes or move them independently." But what difference does it make whether your feet are strong if you can provide them with the support they need? It makes a big difference, according to Davis. "Shoes should not take over the job that the feet are supposed to do, because that means the feet will not develop strong muscles. Weak feet then become dependent on artificial support. And if that support is lacking—shoes deteriorate with wear—then the foot is no longer able to tolerate the normal strain of walking or running, making it more prone to injury."

Most runners like their motion control shoes because they are very comfortable, but Davis still regards the phenomenon as a peculiar aberration. "There isn't a doctor on the planet who would consider fitting a neck injury patient with a permanent brace. A brace is only ever fitted temporarily during rehabilitation, with the ultimate aim being to strengthen the neck and the neck muscles again. The same applies to every other part of the body, except for the feet, it seems. We need to start thinking the same way about our feet, too." She may have a point. If, for instance, a person breaks one of their vertebrae while snowboarding, they have to wear a brace for many months to keep the spine in the correct position. Then the patient has to work very hard to strengthen the weakened muscles again. They also have to be extra careful with their back for the rest of their days and keep the muscles in good shape. So if your back muscles become weak when they are given extra support, why should it be any different for your feet?

According to Davis, we often overlook the foot when treating injuries, and that is why strengthening the feet is a critical aspect of treatment at her center. The leading role is for the arch of the foot, which acts as a shock absorber and spring mechanism. She explains how the arch stretches and bends while we walk. When we are standing still the arch is flat, and when we lift the foot to walk it springs back into its arched position. This ability to flatten out is crucial to distributing the shock of landing, the foot's own form of cushioning. The second task of the arch is to act as a spring. A flattened arch stretches the elastic tissue to absorb the energy of landing, which is released again when the arch springs back into shape with the next step. Davis says that a strong arch forms the "core" of the foot and compares its importance with the core stability of the torso that is essential to good posture.

Davis has not yet been able to prove that shoes cause fallen arches, but she does believe that weak feet lead to problems like heel pain. It is also one of the reasons why she thinks that running barefoot is good, because it keeps the foot muscles strong. A 1987 study showed that walking and running barefoot for four months makes your feet almost one shoe size smaller, as the arch becomes stronger and consequently higher. Davis reaches down and pulls a box from under her desk. "I always run barefoot, but I wear these when the weather is bad or when it's freezing." She extracts something from the box that looks more like a pair of ballet shoes than running shoes and proceeds to bend them almost double. "See how flexible they are?"

Thick soles with cushioning are not her thing either, and not just because they can cause your muscles to turn to jelly. "Studies have clearly shown that a runner's foot lands hard in cushioned shoes." But surely the cushioning should mean a softer landing? Davis uses a boxing glove analogy to explain the effect to me. "To avoid causing yourself an injury, you won't hit an opponent as hard with your bare hand as when you are wearing a boxing glove. A punch from a hand encased in a boxing glove is much more powerful." The same idea applies to shoes, too. "The moment you add cushioning under your heel, the harder you will allow your foot to hit the ground. The sole compensates a little, but the impact with the ground is still considerable, without you really feeling it."

Some human movement scientists view the powerful forces that impact the leg as a kind of wrecking ball that sets to work on our bones and muscles tissue. They believe that the softer the foot lands, the lesser the impact on the knee, lower leg, and heel, and that means fewer injuries. Easier said than done, unfortunately. So what can we do to cushion the blow? After all, whichever way you look at it, running requires you to jump constantly from one foot to the other. According to Davis, the answer is simple: we need to land our feet in a completely different manner. Not on the heel but on the forefoot, and that can be done more easily when you run on bare feet. Many barefoot runners run this way automatically. They land on the ball of the foot, allow the heel to briefly touch the ground, and then push off again. Runners who wear shoes usually hit the ground with the heel first before rolling onto the forefoot. So does this alternative method of landing make barefoot running safer?

HEEL BAD, FOREFOOT GOOD

According to evolutionary biologist Lieberman, the habit of landing on the ball of the foot stems from the way in which prehistoric humans ran after their prey two million years ago. To find out how our ancestors ran, he travelled to Kenya with a number of colleagues. In some parts of the world it is still the norm to walk around barefoot, as the Kalenjin people do in Kenya, a tribe that is also known as "the running tribe." Many of the best Kenyan athletes belong to this ethnic group, including the former world record holder for the 10,000 meters and the marathon, Paul Tergat.

In Kenya, the scientists carried out a detailed analysis of the running techniques of these indigenous people and compared them with the techniques of shoe-wearing runners. The result? People who grow up walking around barefoot land more often on their forefoot, while people who are used to wearing shoes tend to land on their heel. This wasn't exactly earth-shattering news, as it had been observed before. However, Lieberman's team did notice something else that was definitely newsworthy. It was clear that barefoot runners who landed on the forefoot landed softer. Shoe-shod heel-landers experienced a harder impact with the ground with each step, which meant that their legs had to deal with more powerful impact forces. In other words, landing on the forefoot means you can avoid the crushing impact on the leg that many scientists believe causes significant damage to our knees and shinbones. Whether running barefoot actually prevents this abrupt impact is still a matter of discussion. The impact may be more evenly distributed but, in the opinion of some scientists, that does not necessarily mean that the landing is softer.

Let's assume for a moment that running long distances was part of the survival strategy of the first humans. In that case, you would expect natural selection to have done its best to reduce the risk of injury to a minimum. Nature's solution was to develop a running technique that had less of an impact on the body. Lieberman suggests that landing on the forefoot was probably the norm back when humans were running around the savanna on bare feet, but he is not 100 percent sure because he still does not have the kind of irre-futable evidence needed to confirm his hypothesis.

I decided to give it a try. During the analysis at the Catholic University in Leuven, I did a session without shoes on the treadmill as well. It took a little getting used to. I tried to land my feet softly but I felt about as graceful as a bull in a china shop. "I had expected you to switch automatically to landing on your forefoot," said Ellen Maas, who conducted the test. "But you're obviously not used to doing that. This test was probably too short, though, to see if you could switch automatically." It is not unusual to assume that a person will automatically run differently on bare feet. It can be very uncomfortable or even painful to land on an unprotected heel. I expected the worst from the test, but the results weren't as bad as I had feared. The impact shock when running barefoot was actually a little less compared with the test while wearing shoes. There was also very little difference with regard to the angle of my hips and knees—more or less the same as with shoes on. Maas identified another plus: I had more forward thrust and a stronger push-off when running barefoot. "That is good for your running economy because it means you are conserving energy." But I also had less stability when running barefoot. In terms of the impact shock on my legs, my feet were not symmetrical. We concluded that I would be better off not running barefoot, at least not in the way I had done in the test by landing on my heel. All in all, it just didn't really look very fluent.

In general, however, running on bare feet does not form a barrier for athletes in terms of performance, a fact that has been proven by the Ethiopian marathon runner Abebe Bikila and the South African athlete Zola Budd. Both managed to run world records in their bare feet. Bikila did so in the marathon at the 1960 Olympic Games. Back in his homeland he usually wore running shoes because of the muddy and stony conditions of the paths he ran on. But the course in Rome was so level and the surface so smooth that he figured he could run it without shoes. The story goes that when his fellow competitors saw Bikila's bare feet one of them quipped, "Oh well, that's one we can beat." Whether because of his bare feet or not, Bikila set a new world record for the marathon that day. He won the marathon at the next Olympics, too, but this time wearing shoes. Zola Budd grew up in South Africa before moving to Great Britain. She was also well known for running barefoot during training and in competitive races. In 1984, at the age of seventeen, she broke the world record for the 5000 meters at a race in South Africa and broke it again one year later in the UK. Ever since, the taxis in her native land have been called "Zola Budds."

BAREFOOT RUNNING ≠ FOREFOOT LANDING

According to the author Christopher McDougall, if we had grown up without hi-tech shoes, we would probably run on our forefeet like the Kalenjin and the Tarahumara. But because we were raised to wear shoes when running, we are doing it all wrong. Anyone who has read his book *Born to Run* has probably come away feeling that landing on the heel is the enemy and landing on the forefoot is far superior. McDougall suggests that walking in comfortable shoes has changed our way of moving so dramatically that it has become a common source of injury. The theory might appear attractive, but it has a number of flaws. If landing on the forefoot is so superior, why does the Olympic champion Meb Keflezighi set his heel down on the ground first when running? And why does the British athlete Hannah England, who won a silver medal at the World Athletics Championships in 2011, do that too? Not all of the world's leading athletes land on their forefeet. If you watch a marathon on TV, you will see that every runner has their own way of landing. The idea that all top athletes land on their forefeet is a myth that has somehow managed to give this running technique a very good reputation. And the greatest misunderstanding of all is that the two categories are mutually exclusive: that all barefoot runners land on the forefoot and all shoe-wearers land on the heel.

"That is an oversimplification and probably even incorrect," was the opinion of a group of South African and Canadian scientists in a review of barefoot running published in 2014. Among the studies they referenced was one in which landing on the heel was shown to be very common among the Daasanach, a tribe in Northern Kenya. Almost three-quarters of the thirty-eight members of the tribe tested landed on their heel. The faster they ran, however, the greater their tendency to switch to landing on the forefoot, but this certainly did not apply to everyone. Forty percent of the test subjects were consistent heel-landers. There are also differences among shoe-wearers. Nine out of ten recreational runners land on their heel, with the rest landing differently, including on the forefoot and even on the midfoot, which represents a third option. Furthermore, splitting landing habits into three categories might actually be a completely pointless exercise, according to another group of contributors to the *British Journal of Sports Medicine*, given that landing one's foot usually involves a combination of all three.

Barefoot running is therefore not synonymous with forefoot landing. One thing that is unique to barefoot running, however, is feeling the surface you are running on. Some experts believe that if your feet provide you with enough feedback while running, you will be able to choose the best way to land your foot yourself. You adjust your technique according to the information you receive from the sensors in the soles of your feet. For example, you will probably avoid crashing down on your heel while running on asphalt because it's just too painful, but doing so on a sandy beach probably feels pretty comfortable. While the "bare feet versus shoes" debate continues to center around the best way to land your foot, there has been a slight shift in emphasis in sports sciences recently, with some scientists now beginning to find running techniques a more interesting area of research than footwear. That said, the two appear to be inextricably linked to each other.

THIN SOLES

Modern shoes are accused by some of causing us to run in a way that results in more injuries. Many runners land on their heel simply because that is what they have always done and don't even give it a passing thought. But is it possible that runners suffer injuries because landing on the heel is actually bad for you? Should every runner learn to land on the ball of their foot instead? I put the question to Irene Davis. "Changing your running technique is the best option for every injured runner," is her answer. And not only that.

To reduce the impact on the lower limb sufficiently, forefoot landing on its own is not enough, according to an article written by Davis and her colleagues and published in 2016 in *Medicine & Science in Sports & Exercise*. The runner also needs to wear minimalist running shoes without cushioned soles on their feet.

Minimalist shoes are simple, flexible shoes with a thin sole that you can fold almost double, like the pair Davis keeps under her desk. The design is well thought out: the shoes simulate the experience of running barefoot without you cutting your feet to smithereens on little stones or pieces of glass on the ground. Almost all running shoe manufacturers now have a minimalist shoe on the market.

Thin soles are not a recent invention. The minimalist models that you see now in the stores actually have a long history. In times gone by, everyone used to wear simple shoes when running. The Japanese runner Shigeki Tanaka won the Boston marathon in 1951 wearing split-toe shoes based on the traditional Japanese socks that separate the big toe from the others. There is still a subculture of barefoot runners who have a preference for the split-toe shoe.

Experts can't seem to agree on whether running in minimalist shoes is the same as running barefoot. What they do know is that switching from comfortable to simple carries a risk of injury. "You can't expect someone to change from modern to minimalist just like that. They will only end up hurting themselves," says Davis. Her studies have shown that people who land on their bare heel suffer the most from contact with the ground. Barefoot running and landing on your heel is the worst combination possible in terms of impact loading. If you want to become a minimalist runner, you have to switch to landing on your forefoot. And that's not as easy as it sounds; learning a new running technique requires time and training. Davis has developed an intensive program that helps people to change the way they land their foot without hurting themselves. Its effectiveness has yet to be tested, however.

Lieberman and Davis have an obvious preference for forefoot landing. They are both barefoot runners themselves, after all. However, many other scientists doubt the usefulness of switching from landing on the heel to mid- or forefoot landing. The human movement experts Joseph Hamill and Allison Gruber say, for instance, that there is little evidence that such a switch leads to improved performance and fewer injuries. In a recent article in the *Journal of Sport and Health Science*, they questioned the assumption that landing on the forefoot reduces the risk of injury. Based on the relevant literature, they determined that switching from heel to forefoot does not provide the majority of runners with any real benefits: it does not improve your running economy and you do not suffer fewer injuries. Switching can actually result in a runner putting more strain on muscles that are not normally used when running, which in turn can lead to injury. Landing on the mid- or forefoot can certainly help some runners to run more efficiently, according to Hamill and Gruber, but that doesn't mean we should encourage everyone to run that way. There is too little evidence to justify that kind of advice.

EXPLODING CALF MUSCLES

Landing on your forefoot results in significant impact on parts of your leg that you weren't used to using before. "Before you switch, you first need to train your calf muscles,

Achilles tendon, and the arch of your foot and then slowly build up the miles," explains Davis. You actually need to regard yourself as a beginner who is starting all over again. First you should only walk before you start running, and then slowly increase the number of miles you run in order to build up your stamina and strengthen your ankles and feet. Sports physician Steef Bredeweg, who believes that comfort is the only thing that really matters, agrees with Irene Davis. "Running in minimalist shoes? Fine. But you are only looking for trouble if you try to switch too quickly."

One sunny afternoon, while I was busy researching the material for this chapter, I decided, out of pure curiosity, to go for a run and to land on my forefoot instead of my heel. I had just started running again after recovering from an injury to my shin. After three months of rest, I had to start from scratch again. My physiotherapist had already suggested that it might help if I switched from heel to forefoot. He advised me to switch between the two every few minutes to see if I could feel any difference. This was the day to give it a try. I pulled on my running shoes and stepped out the door onto my forefoot. It felt weird at first, but by the time I got to the end of the street I was enjoying it. I felt so light! My feet even seemed to make less noise than usual. After a few minutes I was completely comfortable with the new technique and I ended up jogging for twenty minutes without feeling anything strange. I paid the price the next morning, however. I woke up with very sore calf muscles and came down the stairs like I had climbed Mount Everest the day before. The following day it was even worse.

But even if you train properly when making the switch, are you not just transferring the problem from one spot to another? The story that anthropologist Jeremy DeSilva tells about his brother suggests as much. His brother, a fanatical runner, suffered regularly from plantar fasciitis (an inflammation in the bottom of the foot), knee problems, and shin splints. When he made the switch to minimalist shoes, these injuries disappeared as if by magic. Now, however, his brother suffers regularly from problems with his calves. "Every form of movement has its pros and cons," DeSilva told me. "Each runner has to find their own solution. Do whatever feels comfortable." The scientific literature appears to support DeSilva's opinion. Landing on the heel may result in stress fractures in the shinbones, but landing on the forefoot brings with it an increased risk of toe injuries. Human movement scientists at the UMC (University Medical Center) in Groningen measured an increase in the pressure on the forefeet of female runners who had switched from normal running shoes to minimalist footwear. This may heighten the risk of microfractures in the bones of the midfoot; a case of out of the frying pan and into the fire, you might say. Mitchell Phillips, a rehabilitation specialist and director of the company StrideUK that analyzes running techniques, agrees that changing the way you land your foot probably only results in different kinds of injuries. "Switching from heel to forefoot might reduce the number of knee-related injuries, but it can also result in more problems with the calves," as he explained in an article in the *Guardian*.

HOLY GRAIL

The ACSM holds minimalist shoes in high regard. In 2014, the organization completely revised its recommendations on running shoes. Extra support for the arch of the foot and a cushioned sole that is thicker under the heel are no longer regarded as essential. Which is good news. After all, why would you recommend hi-tech shoes to runners when there is no evidence that they can prevent injury? The most remarkable aspect of the revised guidelines is that runners are now being told to avoid everything that the organization once recommended. A good, safe shoe now is one that is as light as possible, offers no extra stability or support, and does not have a raised heel. A lot more minimalist than the way things were before. Irene Davis shares this opinion, as do many other experts. But where's the proof?

Davis's human movement lab has produced excellent results. Study after study has shown that her test subjects land more softly on their forefeet, a finding that has been replicated by other research teams too. But does this mean that it reduces the runner's risk of injury? This has not yet been examined sufficiently, and until that happens we cannot draw any conclusions. Establishing the risk or otherwise of injury is the kind of result that matters most. When scientists are testing a new form of chemotherapy, they don't just check whether the tumor shrinks or not. What they really want to know is whether the treatment can extend the life span of patients and improve the quality of life.

But what if the hypothesis is correct? What if landing on the forefoot really does lead to fewer injuries because it eliminates the abrupt impact of landing on the heel? This would signify the discovery of the running world's Holy Grail. In theory, the idea that barefoot running can reduce the risk of injury is not all that far-fetched, according to three orthopedists writing in the *Journal of the American Academy of*

Orthopaedic Surgeons in 2016. "Results to date indicate that a change of technique, not shoes, is the best way to reduce the number of running injuries," says author and orthopedic surgeon Jonathan Roth at the Fort Belvoir Community Hospital in Virginia. You can switch from heel to forefoot no matter what kind of shoe you wear, although simple shoes do make it easier. "Runners can focus more on their running technique when wearing minimalist shoes, but not everyone does that and this can result in more injuries," according to Roth. Some people begin running spontaneously on their forefeet when they start wearing shoes without extra cushioning, but this doesn't apply to everyone. In a study carried out by Dan Lieberman in which runners were asked to train for six weeks in minimalist shoes, half of those used to landing on their heel continued to do so.

The conclusion of the three orthopedists was that the popularity of minimalist running is increasing faster than the availability of medical evidence for its presumed benefits. A study in 2014 into barefoot running drew the same conclusion: right now, no one can say how many injuries barefoot runners suffer compared to runners wearing normal footwear. Ultimately, there is no evidence that barefoot running is any safer. But the opposite applies too: there is also no evidence that running on bare feet does not help.

PLATFORMS

Even though the natural running hype is still very much alive and kicking, a different trend is currently making all the headlines. Since their launch onto the market in 2010, running shoes with extremely thick soles have become increasingly popular. Manufacturers have gone back to the idea of adding more material to the soles with the aim of cushioning the impact when the foot hits the ground. There are currently around twenty different types of these so-called maximalist shoes on the market. They are very comfortable, according to one of my running companions who runs in shoes with platform soles made by the French company Hoka One One. With two inches of rubber under the heel and more than one inch under the forefoot, she is almost as tall as I am now in her new shoes. But do thick soles help prevent overuse injuries?

Researchers are only now beginning to gather evidence for the risk of injury associated with this type of shoe. The first publications in scientific journals appeared in 2015, including one written by Christine Pollard, Associate Professor of Kinesiology at Oregon State University. Together with her colleagues, she has been studying the ways in which runners' feet move in shoes with mega-cushioned soles. In one study they asked twenty participants (injury-free and aged between 18 and 45) to complete a test run wearing traditional running shoes and maximalist shoes, and then to do the same test again six weeks later. The researchers measured the extent to which the runners' ankles turned out when pushing off, the force of the impact upon landing, and how quickly the impact spread upward through the rest of the body.

An earlier study by Pollard revealed that runners with platform sole shoes experienced a greater impact force when landing their foot and that the impact also traveled faster through the body. In this study, fifteen female recreational runners were given a biomechanical examination before and after twice completing a 5K run on a treadmill, first wearing a pair of Hokas and then wearing a pair of traditional New Balance shoes. The results showed that they landed harder when wearing thick soles both before and after the run. Pollard also considered the possibility that the runners needed more time to get used to the thicker sole and that they would eventually adjust their technique after wearing the Hokas for a while.

With this in mind, the team built a six-week adjustment period into their next study. During that period, the runners gradually increased the distances they ran in their Hokas. The results were not what the team expected. At the end of the six-week period, the twenty runners, all former heel-landers, were showing the very same tendency to turn their ankles out as they had at the start. Furthermore, the landing impact when wearing the mega-cushioned soles was once again greater than the impact measured for the traditional shoes. And, in an interesting twist, three of the runners had to drop out during the six-week testing period because of blisters, shin pain, and Achilles tendon problems.

The results of the studies carried out to date sometimes contradict each other. A study by a different team revealed that runners who usually wear normal running shoes do not run any differently in terms of biomechanics when they pull on a pair of maximalist shoes. It is possible, of course, that the discrepancy in the results is due to the fact that the different groups of test subjects did not all wear the same brand of shoes.

We are beginning to find out more, from a biomechanical perspective, about how runners land and move when wearing very thick-soled shoes, but we still cannot say whether these shoes lead to more or fewer injuries. Experts believe that what applies to minimalist shoes may apply to maximalist shoes as well: different kinds of injuries, but not fewer injuries, compared to traditional shoes. "We're probably not lowering injury rates with different shoe types, we're just choosing different injuries," said Richard Willy, Assistant Professor of Physiotherapy and Rehabilitation at the University of Montana, in an article in *Runner's World* in 2019 when asked to comment on Pollard's study. Prone to knee injuries? Then it is probably a good idea to avoid Hokas and other thick-soled shoes. Problems with your Achilles tendons or the soles of your feet? Then a thicker barrier between your foot and the ground might do the trick. There is some evidence that maximalist shoes reduce the pressure on the sole of the foot while running.

Thinking about switching to high-cushioned running shoes? Then you should remember that a pair of shoes with a thick lump of rubber in the sole can weigh up to 16 ounces. If you are used to wearing shoes that weigh 8 ounces, it will probably be no fun dragging all that extra weight around.

HEALTH CLAIMS

So which running shoes are the best? Science may never provide us with the answer to this question. To gather hard evidence you would need to carry out controlled clinical tests, and they take time and are very expensive. You would need hundreds of runners, thousands even, all split up into random groups: one group wearing normal shoes, another wearing minimalist shoes, one wearing maximalist shoes, and another one running barefoot. The test subjects would have to be matched as perfectly as possible and cover more or less the same distances. The runners would have to be followed for at least one year, preferably longer, and then the different groups examined in terms of the number of injuries. Only these kinds of studies lead to evidence that is regarded in scientific circles as being of the highest quality. And until someone has carried out such a study—as well as a follow-up study to corroborate the findings—no one can claim that modern running shoes are bad for you and barefoot running much better, or vice versa.

This didn't stop the American firm Vibram from jumping the gun, however. Vibram is known for its FiveFingers running shoes—minimalist shoes with separate toes. More than 70 million people in the United States own a pair. In May 2014, a large group of people (over 150,000) filed a class action lawsuit against Vibram and the false health claims that the company had been making for its footwear. It was claimed that FiveFingers could strengthen the muscles of the foot and lower leg and improve the mobility of the ankles, feet, and toes, as well as the runner's gait. It sounded too good to be true, and the judge thought so too. In its judgment, the court stated that there was no evidence to support the company's claims. Vibram agreed to settle for $3.75 million to avoid going to court, and the claimants were partially reimbursed.

The only apparent benefit of running barefoot was debunked a few years back. Various research teams had concluded that for every four ounces of extra weight per shoe, the wearer used up 1 percent more energy. The assumption therefore was that running in bare feet costs less energy because the runner is lighter. However, this does not appear to be the case. In 2012, human movement scientists at the University of Colorado calculated that barefoot runners used just as much oxygen as runners wearing light shoes, even though you would expect them to use up less energy. The advantage gained from not having to carry the weight of shoes appears to be negated by some other factor, most probably the running technique of barefooters. In a second experiment, the scientists added a strip of lead to the runners' bare feet to make them as heavy as their shoe-shod feet. In that case the runners saved between 3 and 4 percent more energy when wearing shoes than when running barefoot.

In the end, I decided to make the switch. Not in terms of shoes, but in terms of landing. The problems with my shins were giving me the most trouble I had ever experienced in my running career. The situation really couldn't get any worse, so I figured it was worth a try. Having to start from scratch again is the perfect time to make a radical change. At first I alternated between the two: five minutes on the heel, five minutes on the forefoot. Then seven minutes on the forefoot and four on the heel, followed by ten minutes forefoot, three minutes heel. I kept this up until I was able to run five kilometers without getting cramps in my calves. And five quickly became ten. Eighteen months later, I entered my first 10K race since being sidelined by injury. Free of all expectation, I stood at the back of the pack waiting for the starter's gun. I started the race far faster than I had planned and pretty soon the miles were flying by. I found myself running faster and faster without any trouble at all. I crossed the finish line in a personal best of 48 minutes and 43 seconds. In the months that followed, I smashed my records for the 5K and the half marathon. Was this all down to landing on my forefeet? I have no idea. What I do know, however, is that it took only four months for my shin problems to return. Different landing, same old problem.

Time now to forget about footwear and focus our attention elsewhere. After all, we run not only with our feet but with our whole body, and runners come in all shapes and sizes. So is every shape and size suited to running?

TIP

The best advice that science can give us is: if it ain't broke, don't fix it. In 2009, a group of Australian physiologists suggested the following: if you wear modern running shoes and they give you no trouble, great! Don't change anything. On the other hand, if you have no particular problems with your feet but still suffer injuries, consult an expert, like a physiotherapist or an orthopedist, about possible alternatives.

4

BUILT TO RUN

I am a runner because I run. Not because I run fast. Not because I run far. I am a runner because I say I am. And no one can tell me I'm not.

—John Bingham, author and marathon runner

The next time you go to watch a major marathon, have a good look at those taking part. The first runners to cross the finish line will probably be pretty similar to each other in terms of physique: slim, skinny even, and not very tall. Go away and come back two hours later for another look. The difference will be like that between day and night. The runners crossing the line now will have all kinds of bodies: short legs and a large upper body, long legs and a stumpy torso, overweight, underweight, broad shoulders and wasp-like waists, wide hips with bow legs, narrow hips with knock-knees, and so on.

According to Daniel Lieberman, humans are born to run long distances. But does he mean *all* humans? Most of us can run a hundred yards to catch a bus if we need to, but running a full or half marathon is a completely different undertaking. Professor Jamie Timmons at the Loughborough University in England estimates that 20 percent of us will never be able to run a marathon no matter how hard we train. So what makes us able (or unable) to run long distances? Is it simply a matter of physique? Most people guess that I'm a runner when they first meet me. I have attempted to run a marathon a number of times, but each time I was thwarted by injuries. Maybe I'm not built to be a runner and should take up some other sport instead. Or maybe I am a born runner, but over short and not long distances. Time we went in search of the ideal running physique. First up is the question: are women less suited to running than men because of the way their bodies are built?

MANLY WOMEN

Not that long ago it was believed that the female body was not built to run long distances. Up until the 1960s, there was widespread prejudice and misunderstanding regarding the athletic abilities of women, as has been documented in the book *Running across Europe*. People even believed that running was an unhealthy practice for the female body and that overexertion could make it difficult for a woman to have children. The woman who finally rubbished that bizarre idea was an American by the name of Roberta Gibb, who is now in her late seventies. Gibb was born with a desire to run. In the 1960s, when she went to watch the Boston Marathon and found herself fascinated by the event, she decided to start training with a view to entering the race the following year. Unfortunately for Gibb, this was a time when it was considered "improper" for women to run, recreationally or otherwise, and that they should only take exercise in private and never jog anywhere but in their own garden. Better still was to avoid all that effort and sweat and to dedicate one's time to more graceful forms of movement, like dancing. The only thing running did for a woman was to make her more "manly."

The repressive spirit of the era was made all too apparent to Gibb when she tried to register for the race. The director wrote her a letter explaining that she could not enter because she would not be physically able to take part. Gibb wasn't planning on taking no for an answer. But the only option she had left was to enter the race illegally. On April 19, 1966, the 23-year-old Gibb found herself standing in the bushes wearing a hoodie and a pair of her brother's shorts waiting patiently for the start of the Boston Marathon. When the starter's gun was fired, she let the men fly by before tagging along at the end of the bunch. In the end, Gibb, who in the meantime had been spotted by the public and cheered on raucously during the race, finished among the top 30 percent of the runners.

Women were eventually allowed to officially enter the New York City Marathon and the Boston Marathon in 1971. It was not until the 1990s, however, that women started jogging in large numbers. Today everyone agrees that both men and women are capable of running long distances without any trouble. Nevertheless, there is one attribute that can make life as a runner more difficult for women: breasts.

BOUNCING BREASTS

Now that running and taking part in marathons have become popular with women, the female breast is finally receiving the attention it deserves in the world of sports science.

The research team led by Joanna Scurr at Portsmouth University in England, for example, studies breast health and how breasts are supported during physical exercise. Without the proper support, breasts tend to bounce around a lot and cause pain and discomfort, thereby taking the pleasure out of running. Scurr has spent the past few years examining in great detail how breasts behave during running. In one of her studies, women with a size D cup had pieces of reflective tape attached to their nipples and torso so that Scurr could track their movement in 3D using a set of cameras. The women were asked to run on a treadmill, first wearing no top and then wearing a sports bra. What did Scurr discover? Breasts are like ping-pong balls: they move forward and backward, up and down, and from side to side. They display a horizontal figure-eight pattern of movement that becomes larger the faster one runs, up to a speed of around six miles per hour. Above that speed, the pattern remains more or less the same. The breasts do not move in sync with the torso but display a kind of delayed reaction because of their weight.

Even walking without proper support causes the breasts to shift up to two inches—the sum of the movements in all directions—with each step. When someone is running, the displacement can be as much as six inches, particularly in a vertical direction. The breasts still display a figure-eight movement inside a sports bra, but with only half as much displacement.

All that bouncing around stretches the tissue (known as Cooper's ligaments) that keeps the breasts attached to the rib cage and collarbones. If you suffer from painful breasts while running, you are probably putting too much strain on those ligaments. Without the right kind of support they can become damaged, which causes the breasts to sag more quickly. This is not necessarily a disaster, of course, as all women's breasts end up sagging eventually, but I know what I'd do if I found out that wearing a good bra could help slow down the process. Writing in her blog on the RunningPhysio website, the British physiotherapist Jayne Nixon says she can easily pick out women who wear bras that don't offer enough support. They're the ones who don't use their arms while running. Instead of swinging their arms with each step, they keep them down by their sides to hold their breasts in place. And the larger the cup size the worse the bouncing becomes, although this does not mean that women with small breasts do not suffer the same trouble. All that jiggling around can be very uncomfortable for them too.

In another study carried out in 2016, a team at Portsmouth University published the results of a survey of two thousand British teenage girls. Some girls believe that they shouldn't take part in sports because their breasts are too big and are a potential cause of embarrassment. A crying shame, says Scurr. A good sports bra can go a long way toward solving the problem.

What constitutes a good sports bra? There are compression bras that flatten the breasts and bras that encase the breasts in separate built-in cups. There is also a model that combines both of these features. In a survey conducted by Scurr of women with small breasts who took part in the London Marathon, the compression bra emerged as the most popular choice. The built-in cup variety was more popular among older runners. I have both types in my closet at home. I prefer the compression model, a simple top that you can just pull on over your head. I've tested it often enough in front of the mirror and I've never noticed any extreme movement. Perfect for me. However, I usually wear a stronger type of bra when running long distances, one with built-in cups, thick adjustable straps, and two fastenings at the back. It keeps everything firmly in place and is comfortable, too. Which type is better is still anyone's guess, however. Recent studies have shown that the vertical movement of the breasts is equal for both types. What matters is that you find a bra that matches your needs so that you will actually use it when required. At the end of the day, having breasts should not prevent you from enjoying a good run.

KNOCK-KNEES, BOWLEGS, AND FLAT FEET

Another difference between men and women is in the hips. Women tend to have broader hips than men, which means they are more prone to knock-knees. My own hips are broad compared to my waist. But what difference does that make? Scientists at the University of Kentucky recently discovered a link between wide hips and kneecap pain, a complaint that is twice as common among women than men. They analyzed the running techniques of four hundred healthy women over a period of two years. Fifteen of them suffered a knee injury (patellofemoral pain syndrome) during that time. The injured women clearly exhibited more inward movement of the thighs than the rest of the test subjects. "Yes, women have broader hips. But a causal connection between big hips and injuries has not been established," says Ron Diercks over the telephone. Diercks is an orthopedic surgeon and professor of clinical sports medicine at the UMC in Groningen, where he specializes in overuse injuries. He and his colleagues are trying to identify the characteristics that make rookie runners prone to injury. So far, the results haven't revealed anything remarkable. "We have studied the participants in several tests, including the NLstart2run study, from all angles and have seen that women have a higher risk of injury compared to men, but also that it has nothing to do with their hips." The risk of injury is lower for women who are on the pill. "The argument is that ligaments can become looser or tighter because of the hormonal influence of the pill. But that's pure speculation," says Diercks.

I had a long list of physical characteristics, including wide hips, knock-knees, bowlegs, flat feet, and high arches, that I wanted to explore with Diercks, but the professor came quickly to the point. "Scientific studies have never been able to show that physique is an important factor for recreational runners. All kinds of claims have been made, but with little proof to support them. The only link that has been found is the one between knee injuries in female runners and a tendency to turn in one's knees. Probably a little disappointing to hear," he says. No, not at all, I insist. This is good news for me and all the other runners out there. It is also quite surprising. You could be forgiven for assuming that someone with bowlegs automatically puts more pressure on the insides of their knees. Diercks's explanation confounds this supposed logic. But doesn't all that extra pressure resulting from bowlegs lead to more injuries? "If you have a pronounced case of bowlegs, the risk of damage to the inside of the knees is greater anyway because you put more pressure on that area. But that's a long-term effect. If you compare rookie runners with bowlegs with knock-kneed runners, you will see no difference in terms of the number of injuries," explains Diercks. "Most football players have bowlegs, and they can run pretty fast."

There are also plenty of runners, including Olympic record holders, who are flat-footed. An awful lot of research has been done into flat feet and high arches. "No matter how logical it may seem, there is no evidence to show that they have any significant effect in terms of injuries. So there are no real grounds for prescribing insoles or special kinds of shoes, either," says Diercks.

Your body does not have to be perfectly symmetrical to be able to run long distances. Sports scientists believe that the body simply adapts itself as required. If one part of the body is not wholly symmetrical, another part will compensate. And at any rate, we all run asymmetrically, regardless of how symmetrically our body is built. Scientists who ask "perfectly built" runners to run on a treadmill for ten minutes always record great variations in stride length and frequency, as well as in the impact on hips, knees, ankles, and feet. Even with a symmetrical body, every step you take is different. It has been suggested, however, that an increase in asymmetry as a consequence of fatigue can increase the risk of injury.

The NLstart2run study, which was completed in 2015, shows that being overweight does increase the risk of injury for runners. Not all studies connect a high BMI with a higher risk of injury. Indeed, some scientists see no link with weight at all. The difference in results may be due to the fact that some scientists use larger groups of test subjects than others. "Some see a connection, others don't," says Diercks. Not all of the best marathon runners are all skin and bone. The Canadian athlete Peter Maher ran the London marathon in 1991 in two hours and eleven minutes carrying a body-weight of 175 pounds (although he is also over six feet tall).

Anyone who wants to know whether a relatively inactive person can run a marathon should consider the Dutch TV presenter Diederik Jekel. He is living proof that it can be done. In 2014, with the help of a team of sports scientists, he set out to take part in the Amsterdam marathon after only four months of training. Diederik was not exactly what you would call the sporty type before he took on the challenge. He was a little overweight and was fond of a cigarette or two. Four months is extremely short for anyone to prepare for a marathon, let alone someone who couldn't run a hundred yards without having to gasp for breath. And still he succeeded. In a time of five hours and fifteen minutes. On a TV show the day after the marathon, he said—albeit hesitantly—that he was feeling fine. "I've got a sore knee and my muscles are a bit stiff, but that's not bad considering I've just run twenty-six miles."

So what can we conclude? There is no evidence that recreational runners with a certain type of physique are less prone to injury. Neither is there any evidence that because of my wide hips and turned-in knees I would be better off giving up running altogether. Of course, this does not mean that running suits my body, either. If you want to *excel* at a certain sport, however, then the idea of a perfect body does come into play.

DESIGNING AN ATHLETE

For the recreational runner, the ideal body is one that simply stays in one piece while running. For the professional athlete, however, this is not enough. Having the right body for their sport gives an athlete an enormous advantage if they want to make it to the top. In 1988, in a bid to unearth the country's hidden sporting gems, the Australian Institute for Sport began scouring schools for talented athletes between the ages of fourteen and sixteen. Among the institute's discoveries was a tall sixteen-year-old girl, Megan Still, who had the perfect build for a rower. Within two years Megan was rowing at the World Rowing Junior Championships. After nine years of intensive training, she and her crew won gold at the Olympic Games in Atlanta. Before the scouts came to her school, Megan had never even touched an oar, never mind rowed a boat. The first things scouts look for in a possible future champion is the right kind of body. So what do we know about the various body types that are suited to running? In her article "Body of Evidence" in the *Guardian*, the author Vivienne Parry describes the perfect physique for a range of sports. The perfect sprinter is tall and has well-defined muscles and a narrow waist. Sprinters with slender calves and narrow hips enjoy an extra advantage in terms of biomechanics. The perfect build for a marathon runner, on the other hand, is a completely different kettle of fish: light, slightly built, and of average or below average height.

Running a marathon is all about ensuring a good running economy by using as little energy as possible. Athletes with a good running economy use less oxygen than runners with a poor economy running at the same speed. In 1996, the human movement scientist Tim Anderson at California State University in Fresno published a list of physical dimensions and proportions that are most advantageous to the running economy of long distance runners. One of the beneficial characteristics on the list for men is average or slightly below average height. For women, slightly taller than average is said to be better. A low fat percentage is also favorable, as are narrow hips and smaller-than-average feet. The shape of the legs is important too. Long distance runners conserve more energy the closer their muscle mass is to their hips. Slender calves are the absolute runner's dream in this regard. The ideal physique for a marathon runner is that of a so-called ectomorph: someone who is tall, thin, lightly muscled, and has thin wrists and ankles. Mesomorphs, with their well-defined muscles, broad shoulders, and low, narrow waists, also make good long distance runners. Endomorphs, on the other hand, are not your typical marathon runner. They are of stocky build and have short limbs, wide hips, high waists, and small hands and feet. Most people are a combination of all types. Based on the above, I would describe myself as a combination of ectomorph and mesomorph. So could that mean I have the ideal runner's body?

THE IDEAL BODY?

In the run-up to the 2012 Olympic Games, I wrote an article together with a colleague about physique and sport for a popular science website. If your dream is to win a gold medal at the Olympic Games, we suggested, it would be worth your while finding out which sport best suits your body. We went to talk to ex-judoka and sports physician Jessica Gal at her clinic in Amsterdam. After spending years at the top of the judo world, Gal decided to pursue a different career path and now advises top-class athletes on how to optimize their performance. I asked her to run her professional eye over my body and to tell me which sport suited me best.

I am quite tall (5 feet, 10 inches) and have always had a slim build. At 120 pounds, my weight is below average for my height. My hips are wide and my thighs are slightly bow-shaped: they do not rub against each other because of the gap between them. My shoulders are as broad as my hips. I also have small breasts, large feet and hands, and long legs relative to my torso. My body is not overly muscular. "I would say you're more of an endurance runner than a sprinter," says Gal after taking an initial look at me. "And I'd guess you're pretty flexible too." She measured the length of my limbs and checked the suppleness of my joints and muscles. People with very flexible joints—known as hypermobility—run a greater risk of dislocation. Not exactly useful if you intend to specialize in an overarm sport like tennis. However, flexible joints often go hand in hand with limberness, which would make you more suited to a career as a gymnast. After a lot of pushing and pulling, Gal concluded that my muscles were fairly supple, but also that the stability and strength of my left leg in particular was below par. "Fortunately, there are exercises you can do to improve that. You are definitely an endurance athlete, but if I had to choose a sport for you I would say cycling," Gal said to me. "You're a bit too tall for a runner."

FAST-TWITCH AND SLOW-TWITCH MUSCLE FIBERS

Another way sprinters and long distance runners differ, apart from physique, is in the makeup of their muscles. Muscles are made of two main types of fibers known as slow-twitch and fast-twitch fibers, the difference between the two being the speed at which they contract. Slow-twitch muscle fibers contract slowly, as the name suggests, and become tired more slowly too. They get their energy primarily from burning fatty acids, have a rich supply of blood, and contain lots of cellular energy factories called mitochondria. Slow-twitch fibers are red in color because of their high level of myoglobin, a protein with a reddish-brown pigment that stores oxygen in the muscle. Fast-twitch muscle fibers, on the other hand, ensure that the muscle can contract quickly and with great strength; the kind of muscle strength that is needed to be able to sprint or jump. These fibers have a limited blood supply, fewer mitochondria, and are whiter in color because of their lower level of myoglobin. They burn sugars to generate energy. Most skeletal muscles are a combination of both red and white fibers, though they vary in terms of proportion. For example, the muscles that hold your body upright when standing, such as those in your legs and back, contain more slow-twitch fibers because these muscles have to remain active all day long. On the other hand, your arms contain relatively few white fibers so that you can lift heavy objects suddenly if required. There is also a third type of muscle fiber that fits somewhere between slow- and fast-twitch. These intermediary fibers possess both strength and stamina and can burn sugars as well as fats. You can use either strength or endurance training to cause intermediary fibers to behave one way or the other.

Everyone has both slow- and fast-twitch muscle fibers, but the proportions are different for each individual. The makeup of your muscles partly determines whether you are good at activities that require more endurance or more strength. More than half of the fibers in the front thigh muscles of untrained individuals are of the slow-twitch type, but this can vary from as low as 5 to as high as 95 percent.

Biopsies taken from the leg muscles of athletes and examined under a microscope have revealed that marathon runners and other endurance athletes have more slow-twitch than fast-twitch fibers. Some 60 to 70 percent are of the slow type, and this high percentage allows them to consume a lot of oxygen. More than 80 percent of a sprinter's muscle fibers, on the other hand, are of the fast type.

So what about my own muscles? I may have the makeup of a runner, but what kind of runner? I am tall. Sprinters are tall. Is there a sprinter hidden somewhere deep down inside me? "Unfortunately, I cannot tell you what your muscle fiber makeup is," says Gal. "I'd have to take a lump out of one of your muscles for that, and even then the composition can differ depending on the muscle group."

PREORDAINED TALENT FOR RUNNING

Letting someone take a lump out of my leg for the sake of science is a step too far for me. But there is another, albeit less direct, way of finding out what kind of stuff I'm made of. Today a package has arrived in the mail containing a cotton bud and a plastic tube. I open it and scrape the inside of my cheek with the cotton bud, which I then place in the tube and return by post to the DNAfit laboratory in London. There the DNA will be filtered out of my cheek cells and tested in order to create a genetic profile. Humans are almost identical to each other, genetically speaking, but there are slight variations in each of the 20,000 genes in our individual profiles.

It is the combination of all these tiny differences that makes the genome of each person unique. Scientists are trying to establish whether the genetic differences between people can be used to predict athletic performance.

Genes are not the only factor, however. While your DNA may indicate that you are potentially good at endurance sports, other external factors, such as training and nutritional habits, are important in determining the eventual outcome. However, Professor Jamie Timmons at Loughborough University believes that, out of all humans, one in five are unable to run a marathon in a reasonable time because they lack the right combination of genes. Muscles usually react to endurance training by growing tiny blood vessels between the fibers to take in more oxygen. Timmons discovered that the unlucky ones among us have muscles that are unable to do this. And no one can run for four hours without taking extra oxygen on board. No matter how hard these unfortunate runners train, they will never achieve any significant improvement. Timmons developed a genetic test to identify such cases. DNAfit tests a total of forty-five genes that have been linked in scientific journals with sporting ability. I hope that a scan of my genes will tell me more about my own running ability.

THE SPEED GENE

Two genes in particular have been studied extensively and are associated time and time again with athletic performance. The first is ACTN3 (alpha-actinin-3), a gene that is closely linked with strength. It contains the code for making a protein called actinin 3, which is found only in fast-twitch muscle fibers and coordinates contractions by connecting those proteins that play a role in contracting muscles with each other. Actinin enables the muscle fibers to contract in an explosive manner.

Humans have two types of the ACTN3 gene: an active one that produces protein and an inactive one. Each person has two types of each gene: a copy from their father and a copy from their mother. It is possible to have two inactive types of ACTN3, which is the case for 18 percent of all Europeans. Their muscles fibers do not produce any actinin, but this does not present any problems. In 2003, the scientist Nan Yang at the Children's Hospital in Sydney, Australia, and his colleagues at the University of Sydney discovered that only 6 percent of the best sprinters in the world have two "inactive" types of ACTN3. The vast majority of sprinters have at least one and sometimes two active versions of the gene. This probably gives them an advantage in strength- and sprint-related activities. Their finding that the broken version of the gene is underrepresented among sprinters has been corroborated in fifteen subsequent studies. Since then, ACTN3 has become known as "the speed gene."

So if people who produce actinin are good at sprinting, does this automatically mean that those who don't produce it are better at endurance sports? Studies of mice suggest, at first glance, that this is the case. In an article in *Nature Genetics*, the research team at the Children's Hospital in Sydney described an experiment with mice that didn't produce any actinin. They were seen to have a significant advantage when it came to stamina. They were not as strong as their actinin-producing counterparts, but they were able to run for longer on a motorized treadmill without becoming tired.

There is less evidence of this for humans, however. Scientists have found two inactive versions of ACTN3 more often among female endurance athletes than among the general public, but the evidence is not very convincing. The gene appears to be more of a "speed gene" than an "endurance gene." Personally, I don't think I was born to be a female Hercules, so I expect to be told that I have two inactive types of ACTN3.

MORE BLOOD FLOW

A gene that is more closely linked with endurance sports is ACE (angiotensin converting enzyme). You can inherit a short or long version of this gene from your parents. The ACE gene contains the genetic code for a protein that helps regulate blood pressure. In 1998, British physiologists discovered that ACE also plays an important role in determining the stamina levels of muscles. They saw that mountain climbers who successfully ascend peaks above 26,000 feet without the help of extra oxygen often have two long versions of the ACE gene.

The ACE gene is one of the most extensively researched genes in the field of athletic ability. Roughly a quarter of all people have two long versions of the gene, while half have one long and one short version, and the other quarter two short ones. If you have inherited a long version from both your parents, as a runner you can count yourself very lucky. Intensive effort has a greater effect on the stamina levels of people with two long versions. In other words, they are more trainable and make more progress at a faster rate. The human movement scientist Sander van Ginkel discovered precisely this while researching his PhD at the Free University in Amsterdam and the Manchester Metropolitan University. I interviewed him for *Runner's World* magazine and he told me that people with two long ACE genes appear to grow more capillaries around their trained muscles. Capillaries are the smallest blood vessels in the body that can feed individual cells with oxygen and nutrients. The more blood flowing through the muscle fibers, the longer the muscle can carry on contracting without getting tired. "Random surveys of marathon runners or endurance athletes often reveal that they have two long versions of the ACE gene. And you are more likely to find the shorter version in sprinters and weightlifters," according to van Ginkel.

The more you train your muscles for stamina, the more capillaries they will grow. That's a simple fact. The theory is that people with two long ACE genes grow more capillaries, which helps them to improve their performance at a faster rate, and those with two short versions perform better at sports in which sprinting or short, powerful bursts of energy are required. Athletes with one long and one short version of the gene fall somewhere in between, although their physical performance is more similar to that of athletes with two short ACE genes. "If you took a lump out of the leg of an endurance athlete, you would see narrow fibers coated in small blood vessels," explains van Ginkel. "The fibers in the muscles of weightlifters, for example, are usually thick and have fewer blood vessels." It is more difficult, therefore, for an explosive athlete to supply oxygen deep into the muscle. A combination of thick fibers and lots of blood vessels is rare, according to van Ginkel. So, generally speaking, you either train your aerobic capacity to improve your stamina levels or you train your anaerobic capacity to improve your sprinting ability. If I were to hazard a guess, I would say that I have one long and one short version of the ACE gene.

STARTER PACK FOR ENDURANCE SPORT

This morning when I wake up and check my in-box, the first thing I see is the long-awaited email from DNAfit. Tingling with excitement, I open the attached report. Naturally, I will carry on running regardless of the results, but I am very curious to find out what my genes have to say. The first thing I read is that I have two long versions of the ACE gene, a common trait among the best marathon runners. Bingo! That's one box ticked. But what about ACTN3, the gene related to strength and for which I expected the report to show two inactive versions? The results are surprising: I have one active version of this gene, meaning that my fast-twitch muscle fibers contain actinin and are therefore able to contract powerfully. Not as powerfully as people with two active versions, however, a combination usually found in the strongest athletes.

It appears that I am better equipped for strength- and sprint-related activities than I thought. My overall genetic profile, however, shows that I belong more in the endurance sport zone: 69 percent endurance sport versus 31 percent explosive sports. Although the report does not tell me which sports I am good at and which not, it is a relief to know that my body reacts well to endurance training. While it may be stretching it a little to say that I was born to run, it does feel a bit like that. My genetic results also suggest that I have a higher-than-average risk of suffering an injury to my Achilles tendons and knee ligaments. Great. Fortunately, that risk depends on more than just your genes, and you can avoid the negative effects altogether with the right training and plenty of recovery time.

It is true that without the right genetic makeup you have less of a chance of becoming an extraordinary weightlifter or marathon runner. The good news is that it is not as black and white as you might think. Possessing "sprint genes" does not necessarily make you bad at endurance sports. Take the speed gene ACTN3. The different versions explain only 2 to 3 percent of the differences in muscle performance. The same applies to so-called endurance genes, which imply a high level of stamina. Generally speaking, marathon runners are more likely to have the long version of the ACE gene, but this does not apply to runners from Kenya and Ethiopia. Many of the best long-distance runners in the world come from East Africa, but a study carried out by the University of Oxford shows that they do not possess the long version. "A runner with two short versions of the ACE gene can win the marathon too," says van Ginkel. "They may be blessed with other important sport genes." There are over two hundred genes that are known to affect physical performance, and each one makes only a small contribution to an athlete's overall success. And it's not all about DNA, either. Your athletic ability is no more than two-thirds dependent on hereditary factors, according to a study of twins carried out by biological psychologists at the Free University in Amsterdam.

Another reason why I shouldn't take the results of my DNA test too seriously is that nothing is known about the effect of sport genes on "normal" people. The genetic profile that firms like DNAfit put together is based entirely on studies carried out on professional athletes. The fact that recreational runners train far less intensively and in completely different ways may even negate the effect of certain genes. Among those of us who aren't professional athletes, there are also so many differences in age, gender, and ethnical background that you cannot say for sure whether certain genes endow all individuals with the same attributes. However, it is true that genes play a role in the success of professional athletes. In the case of two sprinters who train equally hard and are extremely fit, a genetic variation in ACTN3 can be the difference between qualifying for the Olympic Games and not. For the rest of us, our lifestyle is more of a determining factor than the genes we were born with.

The idea that you could offer athletes a training schedule that matches their genetic make-up, like the one DNAfit provides for a fee, is very attractive. However, most experts believe that we are still in no position to start creating individual training programs based on genetic profiles. Extensive studies have been carried out on the relationship between ACE and ACTN3 and athletic performance. Tests on animals, in which a gene is deactivated and the resulting effect monitored, have also shown how these genes can "tweak" performance. However, we know next to nothing about the mechanisms behind a host of other sports-related genes. Furthermore, it has never been fully established whether people train more effectively and with more reward when their workout is matched to their DNA. The theory sounds great, of course, but does it actually work? "Translating associations found in a population study into individual advice is a very big leap indeed and one that requires all kinds of intermediary studies," said Cecile Janssens, Professor of Translational Epidemiology at the Emory University in Atlanta, in an article in *De Correspondent*. "As of yet, those intermediary studies are either unreliable or have not been carried out, which makes it impossible to fully support the advice."

So is this kind of test useful at all? In practical terms, not really. The research into the effect of genetics on how we react to training is still in its infancy. It is, of course, interesting to gain an insight into (a tiny piece of) your own DNA, if only out of sheer curiosity. From this insight, you can draw conclusions like: "Based on my DNAfit results, it appears that I possess a starter pack for endurance sports." However, any subsequent individual training advice that is based on your genetic profile should be taken with a grain of salt.

CAN YOU BUILD THE PERFECT BODY?

Time to pause for a moment and recap. At this point, I have already established that certain genes and a particular kind of physique can be beneficial to your athletic ability. There is nothing you can do about your genetic makeup, of course, but to what extent can you use training to create the kind of body you want? Not even worth thinking about, apparently. It simply doesn't work that way, explained Peter Hollander, then Professor of Human Movement Sciences at the Free University, in an article in the Dutch newspaper *de Volkskrant*. You cannot aim for a certain type of body by choosing a certain kind of sport. Fanatical marathon runners often have a sinewy, almost emaciated physique. Many swimmers have a broad chest and shoulders. And gymnasts usually look very well-toned in their outfits. However, you won't develop the physique of a marathon runner, swimmer, or gymnast by training for just a few hours a week. Hollander believes that there is a very different reason why some sports go hand in hand with a particular kind of body. In fact, it seems to work the other way around: "Playing basketball doesn't make you taller." But people who are tall are more inclined to play basketball because their physique makes them good at it. Long, tall, skinny Sally will probably score a lot of points playing basketball, and that will motivate her to keep on playing. And what about our emaciated marathon runner? You won't necessarily lose weight just by running, and a few hours of exercise each week is not going to turn you into a champion like Eliud Kipchoge. Marathon runners are skinny because that is the secret to their success. "Each kilogram that a runner has to carry is a burden, so the lower the weight the better the runner," according to Hollander.

There's not much you can do about the way your body is built, it seems. You can't make yourself taller or shorter, in any event, nor can you trim your hips or broaden your shoulders. The only thing you can really tweak is your weight. But what about muscle fibers—can we tinker with them? Muscle biopsies have shown that marathon runners have a higher percentage of slow-twitch fibers. But is that all down to Mother Nature? And does it mean that sprinters are born with lots of fast-twitch fibers? An estimated 45 percent of the variation in muscle fiber type can be attributed to genetics. Only a handful of genes have been identified as being directly related to muscle structure. This means that the type of muscles you have and can develop is more or less fixed. External influences, such as nutrition and physical exercise, also have a say, however, and there is plenty of evidence to suggest that specific kinds of training can be used to change the identity of intermediary fibers.

RESISTANCE TRAINING

Studies of both animals and humans suggest that it is possible to alter the characteristics of muscle fibers through specific training programs. For example, the number of fibers with slow-twitch characteristics (fat-burning) can be increased through aerobic exercise—a 12 percent increase by cycling and 17 percent more for long-distance running. Although it is known that training can be used to alter muscle fibers, the degree to which that is possible is still unclear. Can a person who is born with 40 percent slow-twitch fibers raise that percentage to the level of the best marathon runners, for instance? Furthermore, the frequency, duration, and intensity of the training required to achieve an optimal percentage have yet to be established. Research has shown that it requires an awful lot of endurance training to get fast-twitch muscles to start behaving like their slow-twitch cousins.

As a runner, in addition to running long distances you can also use strength training to stimulate your slow-twitch fibers. The term "resistance training" instantly conjures up the image of bodybuilders pumping iron in the gym, but it does not necessarily have to involve an explosive kind of training. Resistance training, including lifting weights and exercises using your own bodyweight, can help to improve your muscles' stamina, which has obvious benefits for long-distance runners. An important precondition is that you train at a low intensity, meaning lots of repetition and relatively little weight. Personally, I like to do Bodypump training—an hour of weight-assisted squats and lunges. Instead of struggling to lift 100 pounds once or twice, you lift 10 or 20 pounds fifty times in succession. It's not about absolute strength but rather about tensing your muscles for longer periods of time. Athletes who want to improve their sprinting, on the other hand, should avoid low-intensity strength training. They need to train at a high level of intensity, the classic kind of strength training involving little repetition and lots of weight. Olympic weightlifting also falls under this category. High-intensity strength training makes you stronger and helps to increase the percentage of all types of muscle fibers. It guarantees bigger muscles. Physical exercise may not exactly allow you to sculpt your own body, but it does offer the opportunity for some subtle molding.

My search for the ideal runner's body has taught me that physique is very important for those who want to get to the very top in their chosen sport. It is of less consequence, however, for the average recreational runner. Everyone can practice almost any sport they like. You can play basketball even if you are not very tall, and bodybuilders can run a marathon too.

The fact that everyone can run does not mean that it is not a skill that you also need to learn, at least according to the experts. Training your body to get used to the effects of load and the strain of running is more important than having a certain type of physique. Overuse, however, is one of the most common causes of injury. But where should you draw the line?

TIP

All body types are suited to running, so don't let that stop you. A certain kind of physique becomes important only when your dream is to win a medal at the Olympic Games. Female runners would be better off investing in a good sports bra than spending their money on an expensive commercial DNA test. Genetic tests currently offer very little in terms of useful training advice on how to maximize your athletic performance.

5

TRAINING LOAD AND LOAD CAPACITY

The body keeps putting up with your stupidity until it can't anymore.

—Jay Dicharry, *Anatomy for Runners*

If there is one that word that sends shivers down the spines of runners everywhere, it is "overuse." And with good reason. A trip to a physiotherapist or sports physician to report an injury often results in an "overuse" diagnosis and the advice to take a break from running. Overuse is the cause of almost all chronic running injuries and is usually the result of poor training habits: too fast, too far, too often. Overuse: as a diagnosis, I always find it very unsatisfactory. It just sounds too general, and it never tells me what the problem is or, more importantly, how to avoid it in the future.

It is easy enough to explain what overuse means, in theory anyway. Take a look at the (simplified) graph below. The Y-axis shows you how much load your legs have to deal with when running, and the X-axis shows the number of steps, that is, the frequency with which you are exposed to that load. The curved line indicates the threshold value for overuse. The line begins top left and continues like a hockey stick down to the bottom right. You want to avoid crossing the line into the domain of overuse. Anywhere below that line is fine. High stress combined with little repetition is good, as is frequent repetition combined with low stress. But when both repetition and stress rise, you are in danger of crossing over into injury territory. Of course, in reality there is no thin line separating healthy from overused tissue; that would be far too simplistic. The model would actually be a lot more accurate if the line were ten times thicker.

The curve is not set in stone, either. First, bones, muscles, tendons, and ligaments all tolerate stress and recover from it differently, depending on the tissue. For instance, tendons are elastic and can cope well with traction. Bones, on the other hand, are better at dealing with compression. Second, the curve can move upward as you strengthen the various tissues, for example by gradually increasing the distance you run.

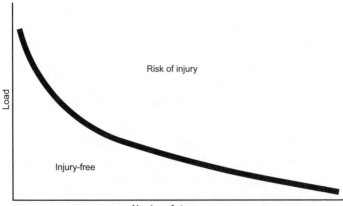

Over a hundred years ago, scientists already knew that organic tissue changed according to the amount of force to which it is subjected. This change happens smoothly when physical exercise is followed by sufficient rest. During an intensive workout, you actually do a lot of damage to your body, but given enough recovery time it bounces back even stronger than before. This is really quite weird when you think of how most machines only deteriorate with wear. Our body is different, however, and when given enough time to adjust, our muscles, tendons, bones, and ligaments become stronger and more efficient. They also become better at coping with a rising training load. It works the other way around, too. When you significantly reduce the stress on a certain muscle, or even eliminate it altogether, the muscle will begin to break down. And that is when the risk of overuse shoots up. If you have ever had your arm in a cast, then you know exactly what inactivity does to your muscles: they disappear. The Dutch astronaut André Kuipers is all too familiar with this phenomenon. When he returned to Earth in 2012 after spending six months in space, he had to be carried from the capsule because his legs were so weak he couldn't walk. The lack of gravity and physical effort in space had drastically reduced his body's muscle mass.

I have suffered more overuse injuries than I care to remember. I am always trying to run faster and farther so I'm an easy target. Of course, almost every runner has the same goal: to ensure that the body can build stronger muscles and tissue at a faster rate than training breaks them down. The question is, how do you balance increasing the amount of physical effort with keeping injury at arm's length?

ADAPTED TO RUNNING

As a beginner, you are not immediately equipped to run for miles and miles. At the Catholic University of Leuven in Belgium, the research team led by Benedicte Vanwanseele studies the behavior of tissue during a beginner's running program. They analyze the Achilles tendon, for example, to see how much thicker and stiffer it becomes and monitor the size and density of the bones. "It takes about three months for the tendons and muscles of a rookie runner to adapt to the stress of running," says Vanwanseele.

Muscles are always the quickest to adapt. Regular resistance training, like lifting weights, makes the individual muscle fibers thicker, which allows you to generate more power. Your muscles do not grow significantly during the first weeks of strength training, but you do become stronger. The reason for this lies in the nervous system. The brain first has to learn how to control the muscles more effectively by activating more muscle fibers at the same time. It is only after six to eight weeks of training that the fibers start to grow. They expand because they start making more proteins, a process explained to me by the muscle physiologist and former decathlete Jo de Ruiter at the Free University in Amsterdam. "Usually there is a balance between the production and breakdown of proteins in the muscle," he tells me. "However, when you work out you disrupt that balance because you stimulate production." After a strenuous sprint or strength training session, the muscles replace damaged proteins with a greater number of fresh proteins. The result? Bigger muscles. "This applies in particular to strength training," says de Ruiter, who is also an experienced athletics trainer. "In the case of endurance sports like running or cycling, the muscles actually become thinner with training, which is more beneficial to endurance athletes because it makes it easier for oxygen to penetrate deep into the muscle fibers." That oxygen is needed to generate the energy required to keep the muscles working.

Muscles react differently to the demands of endurance sports. It is all about improving your stamina, not your strength. Regular endurance training results in an increase in the number of mitochondria—cellular energy factories—in the muscle fibers. Mitochondria use oxygen to produce fuel for the muscles. The best marathon runners often have twice as many of these energy factories in their muscles as nonathletic types. In 2016, a research team at the University of Southern Denmark used muscles biopsies to show how the structure of the mitochondria in endurance athletes is also better. So are they just born with different mitochondria or is the difference down to many hours of training? The team's results pointed to the latter, but the evidence is still inconclusive.

Scientists believe that it takes twice as long to make strong fibers as it does to break them down. "If you train for three months and then stop, after six weeks your tendons and muscles will be back to where they started," says Vanwanseele. Bone takes even longer to grow than tendons and muscles, and the extent to which adults can strengthen their bones by running is still a topic of much discussion. Exercise leads to an increase in mineral density, but that increase is so minimal that experts doubt whether it has any clinical relevance. However, exercise does stop the bones from becoming weaker. Your bone density decreases the older you get, that much we know for sure. And sitting doesn't help either. Mechanical load is required to keep the mineral density of bones at a healthy level, and it can even improve it too. Bones benefit most from exercises in which the muscles pull hard on the bones, such as weightlifting. Activities that result in impact, like running and jumping, are beneficial too, while swimming and skating are less so because they demand little effort from the bones and therefore do not make them stronger.

Another factor that influences the structure of tissue is whether you play sports during your teenage years. Regular exercise during puberty helps to build bone mass. If you maintain this level of physical exercise, you will retain that bone mass, but you will not be able to improve it significantly. "It has been shown that the bone density of women is strongly influenced by how active they were before the age of eighteen. The same applies to tendons," according to Vanwanseele. "You can influence their structure by getting more exercise, but the basis is formed during the teenage years."

Your fitness levels rise at a much faster rate than your muscles and tendons can adapt, and that's why, as a rookie runner, you still can't run a half marathon even after three months of hard training. Your body just won't be ready, even though your heart and lungs may be. "When someone starts running from scratch, after three months they are capable of running around three miles. More adaptation is needed to be able to run longer distances." Vanwanseele believes that there is a limit to how much the various body tissues can adapt. "The bones of a marathon runner are not much different from the bones of someone who can run a half marathon but not a full one."

PUSHING THE LIMITS

If you try to push your body too far, you are only asking for trouble. The first thing that crosses my mind whenever I am confronted with a running injury is the impact of my foot hitting the ground. The harder and faster your foot lands, the greater the impact on your leg. However, overuse is not necessarily the result of impact stress, explains Maarten van der Worp. In his doctoral thesis at Radboud University in the Netherlands, he explored the risk factors associated with running injuries. He now works as a sports physiotherapist at the *Stichting Academie Instituut* in Utrecht. "These days, the focus is mostly on keeping the impact as low as possible. We all seem to assume that the impact with the ground is the source of most of the trouble, and this is primarily the result of the clever advertising practices of the shoe industry. It is amazing how much blame people place on their shoes." In the case of shin splints, it is not difficult to see the link with impact, he argues. But what about problems with the side of your knee or your Achilles tendon? Even the most vivid imagination would have trouble linking these with impact stress. So what lies behind these problems? "Elasticity," says van der Worp. I recently visited his clinic for a gait analysis. He identified at least five areas where I could improve my technique, but the most important was the tendency of my right hip to drop. When your left foot hits the ground and your right hip drops at the same time, the iliotibial tract, a tendon plate on the side of the thigh, becomes stretched more and more with each step you take. In this case, the problem is related to balance issues and possible lack of muscle strength in the hips." The injury I was recovering from had manifested itself as a pain on the outside of my right knee, and wearing shoes with cushioned soles to achieve a softer landing would be no help.

Exactly how overuse manifests itself depends on the part that has been affected, but it usually shows up under the microscope as tiny little tears, or "microtrauma" in medical jargon. In the case of running injuries—apart from sprained ankles and other acute injuries—it is always the result of too much training load, where the damage you cause while taking exercise is greater than the rate of recovery. "In the case of chronic injuries, we have to separate two things: the impact force that a runner exerts on his leg and his capacity to absorb that force," says van der Worp. "If the load occasionally exceeds the capacity, that's not a problem. However, when you push yourself beyond your limits repeatedly, that's when overuse rears its ugly head."

DISTANCE AND SPEED

Danish sports scientists believe there is a difference between extending the distance you run too quickly and increasing your speed at too fast a rate. A few years ago they carried out a study of six frequently occurring injuries among runners who land on their heel: pain at the top of the knee-cap (patellofemoral pain syndrome); pain on the outside of the knee (iliotibial tract syndrome); heel pain (plantar fasciitis); inflammation of the knee tendon (patellar tendinitis); inflammation of the Achilles tendon (Achilles tendinitis); and calf muscle problems. They classified each injury under one of two categories, "too far" or "too fast," based on studies in which runners had answered questions about their training and injuries. The conclusions were published in the *International Journal of Sports Physical Therapy*: injuries to the front and side of the knee are linked to the number of miles per week, while injuries to the calves and bottom of the foot are linked to running speed.

It is not only overly enthusiastic recreational runners with no proper coaching who get injured; the best professional athletes suffer problems too. In fact, the athletes are the ones who most often find themselves teetering on the edge. In 2014, Dennis Kimetto ran a world record for the marathon in Berlin only to be plagued for the next three years by injuries, including to his hamstring. Mo Farah, a four-time Olympic champion, suffered a knee injury during the 10,000 meters at the World Athletic Championships in 2017. At the same event, the hamstring of the fastest man on Earth gave up the ghost; during the final race of his career, Usain Bolt was forced to pull up in the 4 × 100 meters relay.

Overuse means pushing yourself too far, too often, or too fast in training. And things will eventually go wrong when you continue to push yourself too hard. The accumulated stress results in a kind of fatigue that your body tissues simply can't handle anymore. In 2013, van der Worp and his colleagues studied the injuries suffered by a group of women preparing for a 5k and 10K race. Running distances of over twenty miles each week proved to be the biggest source of trouble. Will we ever be able to identify threshold values for overuse for different categories of runners? For example: no more than twenty-five miles a week and no faster than 6 mph for someone with five years' running experience? Van der Worp doesn't think so. Two people with exactly the same training schedules and running experience can still have

completely different capacities for impact stress and prone-
ness to injury. Not surprisingly, therefore, my next question
is: what determines your load capacity?

FLUCTUATING RISK OF INJURY

This is not just another question; it is the key question. It
is not that difficult to form a clear picture of one's training
load, as distance, speed, and duration are easy variables to
measure. Doing so for load capacity, however, is a lot trickier.
One is inclined to think in terms of strong muscles and a
certain type of physique. I have been running since 2006, so
I just assume that my body has had enough time to adapt.
Especially because I also do strength training two or three
times a week to keep my upper body and legs strong. And
still I get injured regularly. There must be a lot more to a runner's load capacity than experience alone. But what?

I have traveled to Rotterdam to put this question to the
human movement scientist Marienke van Middelkoop, who
conducts research into running injuries at the Erasmus University Medical Center. "It is very complex," she tells me.
"Your load capacity has a lot to do with strong muscles and
physique, of course, but it is also affected by sleep, mood,
daily activities, your body clock, catching a cold, etc., etc."
And you also have to consider the fact that your risk of
injury can change at any moment, for example because of
hormonal swings or a change in diet. If you gain a couple of
pounds while on vacation, your load capacity may drop as a
result. Even simply feeling a bit off will make you less able to
take on your usual load. Fatigue does the same; it can disrupt
your coordination or alter the tension levels in your muscles.

It is not known, however, exactly what role these various elements play in determining your load capacity.

According to van der Worp, the focus of research is currently shifting from studying individual risk factors to creating a risk profile: a spectrum containing all of the interrelated risk factors. "It would be ideal if we could create a profile for each injury and translate it accordingly for each individual person." He believes that identifying an individual's load capacity is the way forward. Wouldn't it be wonderful if you could monitor your load capacity using modern technology, like a sports watch or smartphone that alerts you whenever you are in danger of crossing into the overuse zone? An innovation like that could make overuse a thing of the past.

Van Middelkoop would love to invent an app that does just that. In fact, the plans are already in place and she is currently discussing the possibility with the Eindhoven University of Technology and the Free University in Amsterdam. "To make the app, we first need to gather a lot of information from individual runners," she explains. "They would have to keep a record of everything they do in their spare time and of how physically demanding their work is, for example. Collecting this information from a large group of people would allow us to identify specific rhythms and patterns."

Most sports scientists believe that new technologies and big data can help to reduce the number of injuries. Today, runners gather data about their techniques and running habits on a massive scale using portable devices. Sports watches with GPS and real-time feedback on stride frequency are the most popular, but there are also sensors available that can be attached to the back of your shoes or even built into your clothing or footwear to record the impact of each step. If you only want to track your distance and speed during training, then GPS is sufficient. But if you also want to measure the impact forces on your shins or the movement of your joints, your portable device must also be fitted with an accelerometer. The most advanced sensor is the IMU (inertial measurement unit), which contains not only an accelerometer but also a sensor that records changes in the angle of the joints (a gyroscope) as well as an orientation sensor (a magnetometer). Companies like Xsens, Shimmer, RunScribe, and ImeasureU have already introduced these kinds of devices to the market, and plenty more manufacturers are anxious to follow suit.

The scientific community is also very interested in getting its hands on the data that runners collect while training. Up to now, researchers have had to base their studies of injuries on the data from a few hundred participants at most. This is beginning to change. Thanks to portable technologies, they will soon be able to analyze the data of thousands of runners training for a race like the New York Marathon. They will be fed a steady trickle of information on things like distance, speed, and stride frequency. And the more data they collect, the easier it will be to find the links between specific training habits and less common types of injuries, such as lower back pain and pain in the front of the knee (patellar tendinitis). Some injuries currently receive little attention from science because the number of cases is too small to merit investigation, but several studies are now being planned using portable devices to identify the relationship between training load and injuries. The expectation is that large amounts of data will eventually help to pinpoint the causes of injuries, or at least solve part of the puzzle.

What can runners do in the meantime to avoid pushing themselves too far? Right now, the best thing you can do is listen to your own body, according to van Middelkoop and van der Worp. A cliché maybe, they both concede, but still very true. Van der Worp: "If you feel an injury coming on, you should take your foot off the gas immediately. How you deal with problems at the start has a significant bearing on the risk of the injury becoming severe."

CRIPPLED CARTILAGE

The worst kind of injury is one that won't go away. It is often said that running is bad for your knees because it can damage your cartilage, the stuff on the ends of your bones that functions as a buffer and absorbs shocks. You don't feel any pain in the cartilage itself because it does not contain any nerves or blood vessels. You will feel pain, however, when the cartilage is damaged and the bones start to rub against each other. So does running always wear down the cartilage and eventually lead to arthritis? Van Middelkoop is currently studying the link between the two. "Arthritis is the deterioration of the joints," she says. "It is usually associated with old age but it is not exclusive to the elderly. Anyone who has torn their cruciate ligaments usually ends up with arthritis in the knee."

With arthritis, the entire joint is affected when the quality of the cartilage deteriorates. Cartilage does not grow back, so there is no cure for arthritis. However, arthritis doesn't always go unnoticed until all of the cartilage has worn away, van Middelkoop reassures me. There are early signs, too, including stiffness. "If someone complains of pain in the joints, arthritis often shows up on their X-rays. It is almost always detected before it has advanced significantly."

There are many reasons why cartilage becomes thinner and softer, including old age and previous injuries, as well as overuse arising from a difference in length between the two legs. Genetics also plays a role. Risk factors include obesity and demanding physical work. Arthritis is most common among those who were used to lifting heavy weights and bending over a lot in the past, such as construction workers. The good news is that running is not considered a risk factor. "It appears that if you are a moderate runner, it should not be a problem," says van Middelkoop. "Elite athletes and ultrarunners do run a higher risk, however."

Recreational runners suffer less from arthritis of the knee and hip compared with professional athletes and people who spend too much time sitting down, according to a recent meta-analysis in the *Journal of Orthopaedic & Sports Physical Therapy*. An international group of researchers from Spain, Sweden, the United States, and Canada combined the results from seventeen studies covering a total of 115,000 participants in one report. At 13 percent, competitive runners have the highest rate of arthritis. The figure for those who sit for too long is 10 percent, and only 3.5 percent of recreational runners are affected. The data relate to test subjects who had been running for less than fifteen years in total. The reviewers were unable to identify the point at which running becomes bad for the joints. Nor could they pinpoint the causal relationship. The most important message is that, generally speaking, running is not directly associated with arthritis. But what if you already have arthritis? Should you immediately quit running? This question takes up much of van Middelkoop's time. "Doctors don't have an answer yet. But I would say: keep on running as long as it doesn't lead to more problems. There is no evidence that it is bad to carry on running when you have arthritis of the knee." All in all, runners have healthier knees than is often assumed.

SOFT SURFACES

In their efforts to avoid injury, runners maintain a whole range of tricks and habits. For example, when I am recovering from an injury I prefer to run on soft surfaces. But does it help? Running on soft as opposed to hard ground seems like common sense; the relationship between a hard surface like asphalt and major impact on your legs appears, at first glance, to be a no-brainer. Scientists are less certain, however. A hard surface does not necessarily mean a hard impact, as demonstrated recently by scientists at the Shanghai University of Sport. They asked twelve heel-landing athletes to run at 12 km/h on various surfaces: concrete, grass, an athletics track, and a treadmill. They found no differences in the impact on the leg or pressure under the foot. Most studies do record some differences, but they are much smaller than you would expect. In 2013, the University of Sao Paulo published an article in which a group of recreational runners were also asked to run at 12 km/h on grass, concrete, asphalt, and rubber. The force of the impact on the heel when landing on grass was, on average, 13 percent lower than for the other surfaces.

Asphalt is harder than grass, of course. If you drop a bouncy ball on the road, it will bounce much higher than if you drop it on grass. Asphalt absorbs very little of the impact of the ball, meaning that there is plenty of energy left over for it to bounce back. Grass, on the other hand, absorbs much more energy, so that little is left for reuse. This principle also applies to runners, with one crucial difference. We are able to do something that a bouncy ball can't: adjust our body to match the conditions. We do this so well that the difference in the impact on our legs is often close to zero, regardless of whether we are running in the woods or in the urban jungle.

Runners adjust the "stiffness" of their legs extremely quickly and unconsciously to match the surface they are running on. Daniel Ferris at the University of California in Berkeley has studied this phenomenon in depth. This stiffness is different from the decrepit feeling in your muscles the day after you have run a marathon. Runners coordinate the movement of their muscles, tendons, and ligaments to allow their legs to work like elastic bands. Stiffness is actually the force that is required to stretch those elastics. On a soft surface, you increase the stiffness and bend your knees and hips less. On asphalt or concrete, your legs are less stiff and you automatically bend your knees and hips more each time you land.

A small experiment carried out by Ferris involving six test subjects showed that runners actually adjust the stiffness of their legs before they take their first step on a new surface. He believes that each time we are about to take a step, we make use of the feedback we receive on our previous step and the information about the surface that is stored in our brain. This allows us to carry on running in the same manner even when we suddenly turn from the public road onto a path through the woods. However, it does require some extra effort to keep your speed up because on a softer surface your foot stays on the ground three times longer before pushing off again. Soft ground deforms when you step on it and you can only push off again when your foot has stabilized. Doing so on sandy ground requires more muscle power and energy than on concrete.

A little stiffness in the legs actually appears to benefit athletic performance, as it enables you to make better use of your elastic energy and to run more economically. However, it also increases the impact stress on the leg. On the other hand, too little stiffness can lead to tendon and muscle injury. Science is still trying to establish the optimal level of stiffness. The stiffness of the legs when landing differs per person because of our individual anatomies. A club of human movement scientists from the East Carolina University discovered, for instance, that runners with a high arch run with stiffer legs than those with a lower arch.

Up until now, epidemiological studies have not been able to identify a definite link between surfaces and injuries. Grass or asphalt—it doesn't seem to make much difference. "Perception is also important, however," says sports physiotherapist and human movement scientist Maarten van der Worp. "If you think that the surface you run on makes a difference, then it probably does for you." Many running experts believe that alternating the surfaces you run on is the best approach. Variety is the magic word, it seems.

STRIDE FREQUENCY

As with surfaces, there are also many different opinions on stride frequency. The best runners take up to 180 steps per minute, even at low speeds. Amateur runners take 160 to 170 steps at the same speed. Because the best athletes tend to have a shorter stride, experts often encourage runners with a long stride to try running with a higher stride frequency. My physiotherapist advised me to do the same when I was having trouble with my shins and my stride is now much shorter. So is 180 the magic running cadence?

There is no ideal cadence for everyone, according to research carried out at Brigham Young University in Utah. You automatically pick the cadence that suits you the best. Professor of Biomechanics Iain Hunter conducted an experiment with nineteen experienced and fourteen inexperienced runners. He knew that well-trained endurance runners automatically select a stride frequency that is economical for long distances. But can beginners do that too? Or do they need a coach to teach them how? When Hunter asked all of his runners to run for twenty minutes at a speed of their own choosing, he saw that they automatically chose the cadence that suited them best. Each runner was instructed to run using five different stride frequencies on a treadmill, including one they could choose themselves, and Hunter measured the ensuing levels of energy expenditure. He found only one difference: the more experienced runners ran faster than the beginners. But every single runner automatically chose a stride that required the least amount of energy. Hunter concluded that the best stride for a person is the one that comes naturally to them, at least in terms of efficiency. However, he believes that athletes who make a conscious effort to change their stride frequency probably pay a price in terms of their running economy. Another study actually found the opposite, though. Scientists at the Free University in Amsterdam discovered that untrained runners choose an uneconomical stride frequency compared with experienced runners, although their study did have a relatively small number of participants.

The scientific opinion is that everyone has a unique stride frequency based on the length and mass distribution of their legs. My own running technique has changed since I started taking shorter steps. According to my trainer, I used to run "like a deer," with long, elegant strides. He now thinks that my style of running is a bit wooden and that I may be wasting energy because of the way I scuttle along.

Is there anything else we can say about the risk of injury? Well, a systematic review carried out by the University of Wisconsin in 2013 based on ten different articles arrived at the following conclusion. By taking more steps per minute your torso doesn't move up and down as much while you are running. The impact from the ground is also less and your ankles, knees, and hips absorb less energy from the shock. Increasing your stride frequency by 5 percent—for example from 160 to 168—can dramatically reduce the impact on your knees. "Stride frequency is the big hype now," says van der Worp. "I think it can work, but not for everyone. Again, the focus is solely on reducing the shock level. I wouldn't advise someone with an Achilles tendon problem to start taking shorter steps, as that puts more strain on the calves and tendons because you land more often on your midfoot and forefoot. But it could be argued that it does help in the case of shin splints."

CORE STABILITY

My record is four minutes. I'm referring here to *the plank*, the number one trunk-strengthening exercise for runners. Core stability is an essential component of every runner's training and experts always stress the importance of a strong "core." But what does that mean? It usually refers to the ability to control the position and movement of your torso in relation to your pelvis, but it remains a somewhat vague concept. No one knows which muscles in the torso form the core and which role they play with regard to stability. Are the hip muscles part of this core? Some go as far as to include the shoulders. And how do you even measure the stability of a person's torso?

Jay Dicharry, a physiotherapist and author of *Anatomy for Runners*, has called core stability the buzzword of the last few years. Previously, strength training that focused on the torso was prescribed only for those with lower back pain, and it seemed to help. Nowadays, the exercises are performed by people who just want to improve their running. Running literature is currently full of articles recommending core stability exercises as a way of enhancing performance. Surprisingly enough, however, there is no scientific basis to the claim that these exercises can make you a better runner, although some studies have shown a link between poor core stability and injuries. In 2004, Professor of Biomechanics Irene Davis discovered that basketballers and track-and-field athletes who had been sidelined by injury had weaker hip muscles than those who were injury free.

In any event, core stability has nothing to do with performing hundreds of crunches until you have developed a nice six-pack, says Dicharry. Core strength and core stability are two different things. The most important feature of stability is not strength but timing. Your muscles have to be strong enough, but the first thing they have to do is become smarter. Strong muscles are of no help if you don't know how and when to use them. When you are running, the trunk and hip muscles have to spring into action before you land your foot so that your body remains nicely upright when your foot hits the ground. Smart use of your muscles requires the help of the nervous system. Your brain needs to be able to pick up and react to signals quickly and that means working on your coordination, for example by doing balancing exercises. Dicharry recommends standing on one leg for thirty seconds. If that's too easy, try doing it with your eyes closed. This teaches your brain to send signals to the trunk muscles on a continuous basis so that it can remain stable.

Another general guideline for strength training: training with dumbbells or your own bodyweight is better than using a workout machine, according to the experts. A machine dictates the movements you make. All you are required to do is push and pull with all your might, without any real coordination or control. This changes when you start training using your own bodyweight and doing exercises like the hip bridge, which requires you to keep your legs steady when you lift them.

COMPRESSION SOCKS

Who ever thought that knee-high socks would become popular again? People who suffer from thrombosis or varicose veins wear compression socks to prevent or ease the pain of edema (fluid retention) in the lower legs. The socks have been shown to improve blood flow. The pressure exerted by the socks prevents the blood from flowing back into the lower leg and staying there, thereby eliminating the potential for further swelling. Patients are sometimes told to wear these socks after a knee or hip operation. Today they are very popular with athletes, who use them to improve performance, speed up recovery, and prevent injuries. But do they work?

Studies into the effects of compression socks have produced conflicting results. The majority say that they do not improve performance, but there are also a number of positive findings. In 2009, scientists at the University of Erlangen-Nürnberg in Germany found that runners who did not wear the socks became fatigued after thirty-five minutes on a treadmill, while the sock-wearing tests subjects held out for an extra ninety seconds. However, these studies are not always of the highest quality. The number of participants is often lower than twenty, and it is very difficult to conduct blind tests because the runners always know when they are wearing the socks and on which leg. This means that you cannot rule out the possibility of a placebo effect. Runners who have positive expectations of the socks are probably more motivated to perform better at the test. Furthermore, the theoretical model explaining why the socks work is far from watertight. And there are multiple theories, too. One claims that compression reduces the amount of vibration that the leg experiences from hitting the ground, while another says that they help to get rid of waste products more quickly. Whether the effect of the socks is on the blood vessels, the muscles, or something else is still unknown.

And what about injuries? Many runners believe that the socks help to prevent and treat shin splints. With that in mind, I decided to wear a pair for a while, but they didn't help and now lie forgotten in a dark corner of my wardrobe. Very little has been written about the link between socks and injuries. There is some evidence that they help to reduce muscle pain and speed up recovery after a run, but this too could be a placebo effect.

In November 2016, sports physician Wessel Zimmermann gave a lecture on compression socks at the UMC in Utrecht during a symposium on running. He had carried out his own study into the use of sports compression socks by soldiers with lower leg injuries and found no difference in pain levels with or without socks. Nevertheless, almost half of the soldiers said that the socks helped, particularly those with chronic calf injuries.

In a user study carried out on thirty-one athletes with calf problems by Frank Backx at the Rehabilitation and Sports Medicine Department at the UMC in Utrecht, the reaction to compression socks was very positive. The athletes claimed that using the socks helped to alleviate their injuries. "It's not just about the evidence, but also the experience," Backx explains. "Generally speaking, compression socks feel good." Even his own calf problems became less severe when he started wearing them. But he tempers this by pointing out that the socks do not have the same effect on everyone. "I am convinced, however, that they can be a useful aid for many runners."

STRETCHING

At the Phoenix athletics club where I train, dynamic stretching is part of our warm-up routine. First we run at a gentle pace for ten minutes before doing a number of stretching and strength exercises. One of the exercises involves spreading your legs, bending over at the waist, and swinging your arms from left to right. Our trainer calls it "mopping." We then work on our technique and do a few sprints before getting down to some serous running. After interval training on the track, we run a couple of laps and stretch our thighs and hamstrings as part of our cool-down. The question is: is there a scientific basis for including stretching in your training regime?

To start with, there are two types of stretching. With static stretching, the joints remain at the same angle, like when you hold your ankle to stretch your thigh muscles. You stretch the muscle until it begins to feel uncomfortable and then hold that position for a while. The other lighter type is dynamic stretching. This involves holding a certain position for a few seconds without forcing yourself too much, the aim being to alter the perception of stiffness, as Jay Dicharry explains in *Anatomy for Runners*. The idea behind it is that briefly tensing and relaxing the muscles stimulates the nervous system, after which you feel less stiff.

Stretching is a controversial topic in the world of running. A debate was sparked when it became clear that static stretching just before a period of explosive sprint training caused performance to dip spectacularly. So would long-distance runners be better off avoiding it altogether? An abundance of research has been carried out, particularly into stretching as part of the warm-up, and the conclusions are many and varied. One study indicates a spectacularly beneficial effect on performance, the next shows no effect at all, and another finds only negative effects. In 2016, scientists from Flinders University in Australia and Edge Hill University in the UK decided to gather all of the various results for long-distance runners together. In their study, they asked whether stretching—particularly the static kind—was of any benefit. The answer was no.

Static stretching before exercise does not improve your performance. In fact, it can even have the opposite effect. Your running economy can also suffer up to an hour after stretching. Research has shown that stiffer muscles and tendons can transfer more elastic energy, with the result that running subsequently requires less energy (oxygen). Stretching after training doesn't prevent muscle pain, either. It can make tense muscles feel a little less tense, but it has no effect on muscle damage. The reviewers were unable to find any link with injuries. For explosive sports like sprinting and the high jump, the story is more or less the same with regard to injuries: stretching neither alleviates nor aggravates damage. Unfortunately, there hasn't been much research into the effect of dynamic stretching on running. The results for sprinters and high jumpers look promising, but science has been unable to determine whether it has any benefits for long-distance runners. Nevertheless, most running coaches do advocate it as a training method.

Strangely enough, scientists do not know what effect stretching has at the physiological level. It is often assumed that it causes structural change in the muscles by pulling apart the network of collagen fibers in the muscle tissue. The tissue becomes looser and longer as a result. However, in studies in which the muscles were measured, for example using ultrasound, scientists saw no real change in the tendons and muscles of people who had been doing stretching exercises for a number of weeks. What the case might be after months or even years of stretching is still unknown. The accepted theory is that in the case of two months or less of stretching, the effect on the muscles is primarily one of perception. The improved mobility may be because of a higher pain threshold: you can stretch further simply because you can handle more pain. Stretching teaches the brain to tolerate more strain on the muscle and this makes you more willing to stretch even further.

Stretching really has only one obvious effect: it makes you more supple. Temporarily, that is. The increased mobility in the joints lasts for only half an hour, after which you return to being as stiff (or as supple) as you were before. The message in a nutshell? Static stretching during the warm-up is of no benefit to runners, although it has not been proven harmful for long-distance runners (unlike sprinters). Do it in the warm-up, or in the cool-down, if it makes you feel good.

FOAM ROLLERS

I recently bought a foam roller to massage the sides of my thighs. To use it, you press down on the rubber roller with your own body weight to massage the underlying tissue. Foam rollers have become immensely popular and you can buy them at any good sports store. As with stretching, massaging with a foam roller can also briefly increase mobility, albeit temporarily. And, similarly, there is no evidence that rolling affects your performance, either positively or negatively. It can make your muscles feel a little less sore after an intensive workout, but it won't help them to recover more quickly, according to a systematic review published in 2015 by US scientists in *International Journal of Sports Physical Therapy*. Self-massage is a relatively new phenomenon that hasn't yet been subjected to much study, so it is still too early to draw any definite conclusions. It's anybody's guess how you should use the thing properly—how long, how often, how much pressure. Scientists have warned, however, that it can lead to excessive pressure on the bones, muscles, nerves, and blood vessels, but not enough research has been carried out into what kind of damage, if any, this may cause. As long as we don't know what the risks are and how foam rollers should be used, we should be careful how we use them.

AN UNSOLVABLE PUZZLE

Time to take stock. Many of the theories on how running injuries are caused and how they can be prevented fail to pass the scientific test. Most runners have an image of injury that doesn't really add up, if we are to believe the results of a study carried out in 2014 by Bruno Tirotti Saragiotto and his colleagues at the Universidade Cidade in São Paulo. He interviewed ninety-five recreational runners with an average of five and a half years of running experience, of whom 45 percent had suffered an injury in the past. What did they think had caused their injuries? The most common causes cited were too much training, not stretching, no warm-up, lack of strength, wrong shoes, pushing the body too hard, and type of foot. Almost half of these possible causes are not supported by scientific evidence.

One thing we know for certain is that a history of injury increases the likelihood of you running into trouble again. This is confirmed in almost every large-scale clinical study carried out. "It is the eternal risk factor from which we can never escape," says Marienke van Middelkoop at the Erasmus Medical Center. No single risk factor lies at the heart of all running injuries. Not even high impact load. The widely held view nowadays is that the causes of running injuries are complex and varied. Individual factors like age and weight play a role, as do training habits, including the number of sessions per week. Health and lifestyle, such as playing other sports, are part of the equation too. An equation that keeps on getting more and more complicated. Everything is inter-connected, and the importance of each individual factor is different for each runner. It appears to be an unsolvable puzzle.

Even if scientists were able to find one all-important risk factor by pumping zillions of dollars into their research, the clinical relevance of the results would still be unclear. What if they found out that stress fractures are the result of people not drinking enough milk in their younger years? Or that a lack of exercise during puberty increases the risk of runner's knee? None of this is of much use to us runners. Despite the fact that there is a lot we do not know, science is constantly gaining new insights into the art of injury-free running. In his work as a sports physiotherapist, van der Worp says that the scientific literature is always there in the back of his mind. "Imagine that you want to teach someone how to run differently after they have recovered from injury. You really need to be sure that it's going to work. The fact is, however, that there is often more evidence for the effective treatment of one injury than there is for the next. Changing an exist-ing running technique isn't easy—it's a big challenge for the runner. So you don't start unless you are pretty sure of the outcome."

Marienke van Middelkoop is not a quitter, either. She is about to embark on a study at the Erasmus Medical Center

involving 3,500 runners who are taking part in the Rotterdam Marathon, the City-Pier-City run in The Hague, and the
Ladies Run in Rotterdam. Half of the runners will follow a
ten-step plan aimed at preventing injury, and the other half
will train as they please. If the plan works, she will initiate
a follow-up project to see how the knowledge gained can be
made available to as many runners as possible.

TIP

Human movement scientist Marienke van Middelkoop has
one final tip for runners:

> Don't be distracted by the advice you get in shops and
> online. Most of it is well intentioned but useless. Trust your
> intuition; that's usually what works best. So if compression
> socks feel good, wear them. The same goes for stretching and
> running shoes.

6

THE RIGHT FUEL IN THE TANK

Pasta is the one food I can't live without. It's the food I eat to fuel my running.

—*MasterChef* jury member Joe Bastianich

When we think of running fuel, we usually think of pasta. Or carbohydrates in general. Not so long ago, in the run-up to a marathon, runners were typically concerned with only one thing: getting enough carbohydrates on board. A big bowl of spaghetti for breakfast and off you go. These days, the world of sports nutrition is a lot more complicated. Stores offer a bewildering range of gels, bars, and candy for hungry runners, who also like to stock up on bottles of beetroot juice and jars of assorted supplements, which are of course available everywhere too. In addition to carbohydrates, a lot of fuss is also made about getting enough proteins into your system to aid recovery. We hear very little about fats, however, almost as if they are irrelevant for endurance athletes. But is that the case?

Most runners are amateurs, but when it comes to food we try to eat like professionals. I have to confess that I don't pay much attention to how my diet affects my running. I do try to maintain a healthy diet, but I don't make any major adjustments prior to endurance training sessions or even races. I think that sticking to a strict diet is a bit over the top, really. I mean, I'm just running for fun, aren't I? And the great thing about being a long-distance runner is that you can eat whatever you like without putting on the pounds. On the other hand, like all enthusiastic runners I also want to improve my times. So should I follow a specific diet after all? And can you actually improve your performance by eating certain foods and avoiding others, or by adjusting your diet according to your running schedule? Or are we just driving ourselves crazy with all the hype surrounding sports nutrition products? Hopefully, science can provide us with a clearer picture of the best kinds of running food.

ENERGY FOR THE MUSCLES

First, a quick check of the basic facts, which you can find in any biochemistry or exercise physiology book. It's all very simple really. Muscles need energy to be able to contract. The molecule that provides the energy required to get your legs moving is called ATP (adenosine triphosphate). A muscle has only a limited amount of ATP available to it at any given time, and that is why we have to tap into other fuels to source this energy molecule. And the faster we can do that, the faster we can run. The choice of fuel depends on the duration and intensity of the effort involved. We tap into a different energy source when sprinting than when we are running long distances. Physiologists have been able to identify the various energy systems that keep a muscle working. Roughly speaking, there are two options: running on sugars or running on fats.

Imagine you are running a 200 meter race. What actually happens? For the kind of explosive power required, the muscle fibers turn to the fastest source of energy: phosphocreatine. Muscles can produce the most ATP per second from phosphocreatine, and fortunately this fuel is stored in the muscles themselves and is therefore readily available. There is one small problem, however. As with ATP, the supply of phosphocreatine is very limited. It allows you to run for only five or six seconds at maximum speed. This means you also need another source of fuel. After phosphocreatine, glycogen is the next fastest source of ATP. Glycogen is the form in which muscles store glucose, or sugar, for later use. The liver is also able to store glycogen, but it takes longer to release that energy.

Okay, time for a bit of chemistry. When you are running, the muscles convert glycogen into a substance known as pyruvate. This process never uses any oxygen, even though sprinters have plenty of oxygen running through their muscles. Pyruvate is then processed in one of two ways. The first involves converting it into lactate, which also requires zero oxygen. The pyruvate is converted first into lactic acid, but that breaks down immediately to form lactate. Hard-working muscle fibers can keep going for about three minutes on this process of converting glycogen into lactate. However, if you want to run more than a few hundred meters, pyruvate is sent instead to the cells' energy factories (mitochondria), where ATP is produced through a series of chemical reactions. This process does involve the use of oxygen. The route that pyruvate takes through the mitochondria is a lot longer than the one that results in lactate, with the result that the speed at which you can run decreases.

Many runners believe that lactate is a waste product that builds up in the muscles. This is only partly true. Only a small portion of the lactate produced accumulates in the muscles. Most of it is recycled as an energy source. Muscle fibers are able to produce ATP from lactate by transporting it to the mitochondria. This is a very efficient process and, in theory, it can keep you running for ninety minutes until all of the glycogen has been used up.

26 MARATHONS

What if you want to run longer distances or are training for a marathon? Our muscles and liver contain glycogen that we can tap into, but it is never enough. If you want to be able to run 26 miles, the muscles need an alternative source. The burning of fats is essential to long-distance running. When we exercise, the fat in our fatty tissue gets broken down and the fatty acids released are transferred to the muscles. There they go straight to the mitochondria, where they are used to make ATP. A man weighing 150 pounds and with a muscle mass of 60 pounds can, in theory, produce enough ATP to run 26 marathons in succession. Drawing only on his supply of glycogen—approximately 2,000 kilocalories—the same person would be able to run no more than two-thirds of a single marathon, because he would have used up his entire reserves of sugar after 82 minutes.

We all have fat, even the thinnest specimens among us, and we can generate a lot of energy from it. The production process involved, however, is slow. You can produce five times as much ATP per second using muscle glycogen than you can using fatty acids. It takes ten times longer to burn fat than it does to convert phosphocreatine, but burning fat is ideal when you want to run continuously for a few hours.

Nevertheless, you cannot run a marathon on fat alone. It would take even the best-trained athlete at least six hours to extract all of their ATP from fats, simply because burning fatty acids is so time-consuming. When running long distances, we burn both fat and sugar, although one energy system is usually more dominant than the other. When you set out, you are running primarily on glycogen. The sugar supply depletes slowly as you run, which causes your blood sugar level to drop. Your body reacts by releasing fatty acids from your fatty tissue, and the subsequent burning of those fatty acids suspends the burning of sugar. But you don't suddenly switch to burning fats when the sugar has run out. Instead, some of the sugar is conserved so that you will have enough energy left over at the end of a demanding period of physical exertion. This is why you are able to break into a sprint even after running 26 miles. When running a marathon, the very best athletes burn approximately the same amount of glycogen as fatty acids in order to be able to continue running at a speed of 12 mph. The shorter and more intensive the race, the lower the contribution that fat-burning makes to the total energy production.

That concludes our chemistry lesson for now. Next: what is the relationship between our muscles' energy requirements and what we put into our mouths?

FEEDING THE ATHLETE

There is no getting around the fact that if you run a lot you need lots of energy. But when does *what* you eat begin to make a real difference? The answer is much the same whether you are an amateur runner or a professional athlete, according to Miriam van Reijen, a triathlete and author of a runner's cookbook. What professional athletes and recreational runners who are training intensively for a race benefit most from, according to international dietary guidelines, is simply "good food": plenty of fruit and vegetables, whole wheat pasta and bread, and brown rice, supplemented with plant and animal proteins (fish, meat, eggs, pulses, and nuts), modest amounts of dairy products, and preferably unsaturated fats, like olive oil, instead of the saturated kind, like butter. This doesn't sound too bad to me. There is no need to follow a special diet, and most sports nutrition experts are immune to the hype that accompanies products like chia seeds, goji berries, spirulina, hemp seeds, wheatgrass, and coconut oil. Their advice in a nutshell: just eat normal, healthy food, and a little more healthy when required. My own diet seems to fit between the parameters of what constitutes healthy food. So does that mean I'm doing everything right? Or would I be better advised, as a keen runner, to add something extra, too?

I decided to visit the sports dietician Jenny Hofstede at the University Medical Center in Utrecht. In addition to her work in the hospital, she also runs a clinic where she advises athletes on matters relating to nutrition. To do so, she splits athletes into different categories. Recreational athletes are those who exercise for one or two hours a week. All they really need to focus on is eating good, healthy food. For the more dedicated and competitive types—those who train three to seven hours a week and for more than one hour per training session—the food they eat can make a big difference. Certain supplements can also play a significant role. The third category is professional athletes who train up to six hours a day. For them, paying attention to what they eat is an absolute *must*.

ENERGY AND RECOVERY

Dedicated runners do not necessarily need to use supplements, but more about that later on. However, according to Hofstede, you do need to pay attention to what you eat if you don't want your food intake to adversely affect your performance. "Taking care of what you eat can prevent energy deficiency and aid recovery—the two most important factors when it comes to improving performance," she explains. Whether what you eat can really make a difference depends on your goal. Many runners simply want to improve their times and participate in running events. "If you race for several days in a row without eating the right kinds of recovery foods in between, your performance may drop from one day to the next. If you don't consume enough carbohydrates, for instance, you won't have enough energy to take part in your second or third race. These are the moments when recreational runners experience the effects of poor nutrition. However, if you are not too worried about improving your times, and your diet is within the normal parameters, then you probably won't experience any problems."

In the case of dedicated and competitive runners, Hofstede focuses on the meals eaten around training sessions:

preferably not too much food beforehand, but still enough carbohydrates to be able to complete a long run. And if the session lasts for a few hours? Then it is important to take carbohydrates on board along the way too. A recovery meal is also recommended after training. Paying attention to these three aspects falls under the category of "sport-specific nutrition."

A recovery meal is one that contains enough proteins and carbohydrates. But what are the actual benefits? To find out, I phoned the human movement scientist Dionne Noord-hof when she was carrying out research into sports nutrition at the Free University in Amsterdam. "Carbohydrates, like those in bread, pasta, and rice, are vital when it comes to replenishing the energy reserves. After a strenuous workout, the proteins found in meat, nuts, and pulses are crucial to muscle recovery," she says. For healthy people who do not take regular exercise, the US National Academy of Medicine recommends eating 0.8 grams of protein per kilogram of body weight per day. So if you weigh 80 kilograms (175 pounds), you should eat 64 grams of protein each day, preferably in portions of 10 to 20 grams. A 150 gram tub of low-fat quark (a spoonable soft cheese popular in Europe) contains 13 grams of protein, while an omelet made using two eggs is good for 16 grams and a steak up to 25 grams. The easiest way is to spread your protein intake over breakfast, lunch, and dinner. Vegans and vegetarians need to eat 25 percent more proteins to stock up on essential amino acids. Well-trained endurance athletes who train intensively have to go one step further: the American College of Sports Medicine recommends they consume 1.3 grams of protein per kilo per day—over 50 percent more than the average person's intake.

Last year, scientists writing in the journal *PLOS ONE* said that 1.8 grams would be even better. There are also guidelines for food consumption immediately after training, says Noordhof. "Athletes should eat 0.25 to 0.3 grams of protein per kilogram of bodyweight within one hour of exercise."

Runners often drink chocolate milk after a heavy training session because it contains both protein and carbohydrates. A large glass will give you 10 grams of protein and 36 grams of sugar. Hofstede always recommends the low-fat kind because you don't need any extra fats immediately after training. It takes longer to digest fats, which delays the absorption of carbohydrates. The same applies to fiber: you can eat it all day, except for directly after training, as it also delays absorption. You should also avoid fatty foods before a race because they can cause stomach problems and may also cause you to skip a high-carb meal and so miss out on the energy it provides.

Furthermore, a fat-free diet is not recommended for athletes, says Noordhof. "It's not a good idea for anyone, actually, because we need fat to be able to absorb vitamins A, D, E, and K." Fats also perform many other essential tasks in our bodies. For example, they help maintain a healthy cell membrane and are involved in the production of sex hormones like testosterone, progesterone, and estrogen. For female runners, too little fat can even lead to lower-leg injuries. A club of American scientists from the University at Buffalo in New York asked 86 professional female athletes to participate in a survey of their eating habits. It showed that the women who had suffered injury in the previous year ate, on average, 63 grams of fat per day, while those who were injury free consumed an average of 80 grams. The injury-prone runners also sourced less of their energy from fats. The National Academy of Medicine recommends sourcing a minimum of 25 and a maximum of 35 percent of your daily calorie intake from fats. For men, who require an average of 2,500 kcal per day, this means a maximum of 111 grams. For women on a daily intake of 2,000 kcal, the maximum is 89 grams. The best fats are the unsaturated kind found in olive oil and nuts, for example.

CARB LOADING

I ran my best-ever time for a 10K on a breakfast of yogurt with fruit and muesli and two slices of chocolate cake, topped up with a chocolate biscuit just before I set off. I did everything exactly as I *shouldn't* have: too much fiber, fat, and dairy—a surefire recipe for tummy trouble. If it had been my intention to improve my time, I wouldn't have eaten the chocolate. As I sat there eating my breakfast, I overheard some of the other runners discussing what to eat and drink and what to avoid. White bread with jam or raisin bread? I ran a good time that day, but not because of the chocolate, of course (though I would like to believe so). I was probably just more relaxed than usual because I had no expectations. Apparently, not being so hung up about my food intake and what I should do in preparation for the race worked to my advantage on the day.

The only time I pay close attention to what I eat is when I am about to run a half marathon. The evening before, I always eat pasta to boost my supply of glycogen. Endurance athletes refer to this as "carb-loading." But how much sugar can you actually store in your reserves? A very fit runner, who hasn't trained for two days but has eaten enough carbohydrates, can store a maximum of 700 grams of glycogen in their muscles and liver. On training days, the average is around 420 grams. Based on a normal diet, the average untrained person can store around 280 grams in their muscles and 100 grams in the liver. For the sake of simplicity, let's assume that a reasonably well-trained recreational runner will have a supply of 400 grams of glycogen available after carb-loading. One gram of carbohydrates provides 4 kcal of energy, so 400 grams gives you a reserve of 1600 kcal. How long will this last? That depends on your body weight, among other things. For each kilometer you run, you use up the equivalent of your body weight in kcal. I weigh 55 kilos (120 lbs), so to run a half marathon I need $55 \times 21.1 = 1160.5$ kcal. In theory, I should manage just fine on my reserves of sugar, as long as I have topped them up first.

So when should you start loading up with carbs to ensure that you are as "full" as possible when standing at the starting line? The advice for well-trained runners is to start eating more pasta, white bread, and sweet jams and jellies three days before a (half) marathon. Between 7 and 10 grams of carbohydrates per kilogram (3.5 and 4.5 grams per pound) of body weight to be precise. "Eating high-carb meals and snacks in the days before a race helps to boost the supply of glycogen," according to Noordhof. "This has a positive effect on performance when the race lasts longer than ninety minutes." The amount you need to consume also depends on your athletic ability. "Someone who is planning to run a marathon relatively slowly does not need to load up on carbohydrates as much as a runner aiming to beat their personal best. Conversely, if you want to perform at your peak, you need to burn carbohydrates instead of fats."

To prevent the supply of glycogen from running dry after ninety minutes, long-distance runners eat and drink lots of carbohydrates during races, too. You don't need any extra energy for a race that lasts less than an hour. However, for a race that takes between one and two and a half hours to complete, you need to top up your energy supply, for instance with energy drinks. If you don't, you run the risk of "hitting the wall"—becoming completely exhausted because you don't have enough glycogen to maintain your pace. Fortunately, I have never hit the wall myself, but I have it on good authority that it's not a pleasant experience.

The general guideline is 30 to 60 grams of carbohydrates per hour. "Previously we believed that 60 grams of glucose per hour was the maximum amount of carbohydrates that athletes could store and use at any one time," explains Noordhof. Glucose is the sugar found in most food and drinks. "However, recent studies have shown that this can rise to as high as 90 grams of carbohydrates per hour when athletes also take fructose on board, through drinks and gels, in addition to glucose." For well-trained athletes who run for longer than two and a half hours, the recommended intake is 60 grams of glucose and 30 grams of fructose—ready-mixed and easy to consume—per hour. The experts warn us, however, that you need to train yourself first to be able to take large amounts of carbohydrates on board during strenuous physical activity. If you don't, you may end up suffering terrible stomach problems.

HIGH-FAT DIET

The better you are trained to run long distances, the more energy you will be able to extract from fat. For high-intensity training, it's the other way around: it makes you better at using glucose. So does that mean that if you want to burn more fat during a half marathon you should eat more high-fat products instead of, say, pasta? Experiments seem to indicate otherwise. It will not increase the amount of free fatty acids in your blood. One thing that does stimulate the burning of fat, however, is fasting. Fasting before a race leads to more fatty acids in your blood. And the longer you fast, the greater the number. This is not a good way to prepare for a race, however. In fact, it has a negative effect on performance.

There is a way in which athletes can make themselves less dependent on sugar: by sticking to a high-fat diet. The South African human movement scientist Tim Noakes is a passionate advocate of this diet. By eating lots of fats and very little carbohydrates for a couple of weeks, you can force your body to start making ketones: naturally occurring chemicals in the body that are produced when fats are broken down. In addition to sugars and fats, cells can also burn ketones as an alternative means of releasing energy from food. Ketones usually appear on the scene only when the body is undernourished or exhausted as a result of physical exertion.

For over thirty years, Noakes sang the praises of a high-carb diet, as described in his celebrated book *Lore of Running*. He has since made a complete U-turn and is now trying to ascertain whether a high-fat diet is as good as a high-carb one, and possibly even better. He says that the brain and muscles can easily access the energy available in sugar, but also that the supply of glycogen is never large enough for the requirements of endurance sports. His hypothesis is that endurance athletes on a ketogenic diet are better at sourcing energy from fats, but more study is required before this can be confirmed. While it appears that a ketogenic diet can lead to an increase in the burning of fats, there is no evidence of any subsequent improvement in performance. There has also been a lot of criticism of the theory from within the scientific community. One sports nutrition expert, Fred Brouns from the University of Maastricht, is not convinced of its merits in any case, and Dionne Noordhof doesn't believe either that it boosts performance. In their eyes, carbohydrates still reign supreme when it comes to endurance sports.

LOSING WEIGHT

Up until now, we have focused primarily on how dedicated runners manage their diets to aid recovery and improve performance. But what if you have started running just to lose weight? Unfortunately, science has shown that taking up a sport like running is not an effective strategy for people who want to shed a few pounds. Physical exercise is very good for your health, of course, but it is not a quick fix for unwanted flab.

The evidence to date suggests that exercise alone is not very effective when you are trying to lose weight. You will shed only a few pounds more than other people who are also slimming but by some means other than taking exercise. We're not talking about running marathons here, but about walking, jogging, and aerobics. The results are much different, however, when exercise is combined with a change in diet. Reducing the intake of calories leads to significant weight loss—in the short term. What happens in the long term is a different matter entirely. People often lose weight in the short term only to pile it back on later. Without further research into the long-term effects, we cannot say exactly how exercise affects weight.

A study published recently in the journal *Current Biology* explains why dieting is a more effective way of losing weight than exercise. One obvious reason is that taking exercise makes you hungry and so you will be inclined to eat more. Add in all the gels and energy drinks that inevitably go with it and *voilà*. Herman Pontzer, who studies human energy expenditure at the City University in New York, doesn't believe it's that simple, however. He and his international team of researchers recorded the physical movement and resulting energy use of three hundred test subjects over the course of one week. You would expect the results to show that the more active the person, the more energy they used. But no. The test subjects who were moderately active did use a little more energy than those who were inactive: about 200 kcal. However, the results showed that the very active subjects did not use more kilocalories than their moderately active counterparts. It appears that the total energy expenditure levels off at a certain point when taking exercise. The actual exercise itself costs extra energy, but the amount of energy that the body has to spend on other activities—metabolism, general bodily maintenance, and sitting instead of standing—decreases. This means that the body adapts itself by spending less energy on other activities, which allows it to maintain a more or less constant level of energy use. It makes sense when you think about it. After a heavy training session, I usually collapse on the couch and stay there for the rest of the evening. When I do so, I am unconsciously compensating for all the energy I have used up by doing as little as possible afterward. Pontzer's conclusion is that getting more exercise is not the key to losing weight.

There is a trick you can use to boost your energy expenditure when you are at rest: high-intensity exercise—a workout that makes your heart pound in your chest and the sweat drip from your forehead. Not a relaxed jog in the park, in other words. When you cycle as fast as you can for forty-five minutes, your metabolism will be higher than normal for the next fourteen hours, according to the results of tests carried out on a group of males by scientists at the University of North Carolina. The cycling test itself cost the test subjects 520 kcal of energy (a 100 gram bar of chocolate or a plate of spaghetti bolognese). In the following fourteen hours, their energy use while at rest was 190 kcal higher than normal. Not exactly mind-blowing, in all honesty. The extra combustion is the equivalent of two slices of brown bread. Without the jam.

What if you want to improve your performance *and* lose weight? Is that impossible? Not at all, says sport dietician Jenny Hofstede. "As long as you have enough energy when you are training. There's nothing wrong with an energy dip at the end of the day. It's all about planning your food intake intelligently, and enlisting the services of a dietician can be a great help."

POWDERS AND PILLS

Time to take a look at the world of supplements. Sport-specific nutrition refers to normal foodstuffs that are used to aid recovery and restore energy, such as an omelet or a granola bar. Sports supplements are manufactured purely with performance in mind and for boosting both recovery and energy levels. They are meant as a "top-up" for your daily nutrition. Gels, energy bars, energy drinks, and protein powder all fall under the category of supplements.

After an intense bout of strength training, such as a Bodypump session, I often drink a protein shake. With the five running and strength training sessions I do each week, I figure the shakes must be of some help with regard to muscle recovery. I don't eat any meat and not much fish, so I have to source my protein elsewhere. I eat lots of pulses, tofu, quark, eggs, and nuts, and while I may not strictly need the protein in powder form, I like the shake for its ease of use; gulping my proteins down as a drink after training means that recovery can kick in straightaway. But what does sports dietician Jenny Hofstede think? "Explosive athletes use protein powder more than long-distance runners simply because they need more proteins. But ease of use plays an important role too. You can get the same amount out of a bowl of quark or a breast of chicken, they just take up more time." Nevertheless, she sometimes advises athletes to use protein powder because you can consume it immediately after training. "Muscle pain as a result of intense strength training can hinder your running the next day. In that case it can be wise to use protein powder, although you could just eat a protein-rich meal instead, as long as you can do so within an hour of your training session." Protein powder is still not as popular among endurance athletes as packets and bottles containing concentrated carbohydrates. These days, runners are almost always handed an energy drink as soon as they cross the finish line. Even after a 5K run. The range of high-carb sport snacks is endless. I can spend several minutes staring at the shelves in a store and still have no idea which one to choose.

For most of us, energy drinks, gels, and energy bars provide extra calories that we simply don't need. If you go for a six-mile run, there is no need to supplement your energy level. It's a different story, however, for longer distances and more intensive effort. But the question in that case is: are supplements better than "ordinary" high-carb food? In terms of ease of use, yes. It's a lot easier to carry and eat a gel than it is to squeeze a sandwich into your pocket and try to gobble it down while running. Then again, during a run you can always pick up a banana from a water station, and that's just as good as eating a gel. A 60 ml gel made by SIS (Science in Sport) gives you 87 kcal. It contains 22 grams of carbohydrates, but no fat or protein, and the amount of salt is negligible. A small banana (100 g) contains 19 grams of carbohydrates and provides 88 kcal of energy. It also contains a small amount of protein, a trace of fat, and a host of vitamins and minerals. So a banana and a gel provide similar amounts of carbohydrates and energy. Are there any significant differences? Yes, the carbohydrates in the gel are all of the maltodextrin kind: short, easily digestible chains of glucose molecules. Only 0.6 g of the 22 grams of carbohydrates in a gel is sugar. The 19 grams of carbohydrates in a banana are almost all sugar, namely, fructose and glucose. According to experts, glucose, fructose, and maltodextrins are ideal carbohydrates for sports, although glucose and maltodextrin can be digested more quickly than fructose. However, when I run a half marathon I always opt for the gel because I can eat it whenever I want. But if you don't fancy the chemical taste of sticky lemon gels, then you'll just have to wait until the next banana comes along.

CAFFEINE

Coffee is not a supplement, but gels and candy to which caffeine has been added are. Caffeine boosts performance, that much we know. How it does that is less certain, but it probably involves several different processes. Caffeine is known to interact with the chemical adenosine. When you are awake and alert, the level of adenosine in your brain cells is low. The longer you stay awake, however, the more adenosine begins to build up in the cells, which makes you sleepy. Caffeine blocks the adenosine by binding itself to the same receptor, thereby preventing the adenosine from activating that receptor. The result? You become more alert, or at least not sleepy.

In 2009, human movement scientists at the University of Connecticut carried out an overview of all of the studies into the effect of caffeine in endurance sports. "The intake of caffeine can be an effective means for endurance athletes to improve performance, as long as it is taken before and/or during exercise in moderate amounts," they wrote in their conclusion. Three to six milligrams per kilogram of body weight can boost performance by 3 to 4 percent. For me that would mean a dosage of between 165 and 330 milligrams, consumed over the course of a race. A lower amount would have no effect at all, while more could lead to stomach problems.

A cup of filtered coffee contains 85 mg of caffeine and an espresso 65 mg. A can of energy drink contains 80 mg of caffeine, meaning that I would have to drink up to four cans of that sickly sweet stuff to achieve the optimal dosage. Some of the GO gels made by SIS contain 75 mg of caffeine. Caffeine is absorbed into the digestive system quite quickly—within thirty minutes it can be found in your blood. Studies also show that it is more effective to take caffeine in capsule form than in a drink or gel. Athlete and author Miriam van Reijen advises marathon runners to take a shot of caffeine just before they set off, as it will reach its peak in terms of effect forty-five minutes later. And to get the most out of your caffeine shots during a race, you should avoid all caffeine the week before. Depriving your body of the stuff means it will get an even bigger boost when you start taking it again.

We are not all equally sensitive to the effects of a caffeine-enriched piece of candy. Different people react differently, partly because of genetics. The most-cited gene in this regard is CYP1A2, which contains the code for an enzyme that breaks down caffeine in the liver. We are all either slow or fast at breaking down caffeine, depending on the variant of CYP1A2 we possess. According to my DNAfit report, I belong to the slow category. Some studies have shown that people with the "fast" variant benefit more from caffeine during exercise. Bummer.

BEETROOT JUICE

No one can write a chapter about sports nutrition without dedicating a few lines to the endless discussion surrounding beetroot juice. One of my running companions at my club swears by it and will do anything to get his hands on the stuff. A few years ago he signed up to take part in a marathon in Athens, Greece. He bought a bottle of beetroot juice with the intention of drinking it all the day before the race. Imagine the panic when he arrived at his destination only to discover that he had left his precious cargo at home. All he could do was run to the nearest supermarket, where he found no juice but only raw beetroot instead. Feeling a little relieved, he forced down half a beetroot before the race. When he got back home to the Netherlands he told me that it had tasted even worse than the bottled stuff. But at least it had calmed him down. I never drink beetroot juice because the taste reminds me too much of the soil you use to pot plants (not that I am in the habit of eating that, either). However, there is no shortage of endurance athletes, both amateur and professional, who like to experiment with it. So what do we know about beetroot juice?

The active ingredient in beetroot juice at the center of the debate is nitrate. It enters the blood through the gut and then finds its way back to the mouth where it is converted by bacteria into nitrite, which you then swallow again. The body converts the nitrite into nitric oxide, which plays an important role in a number of bodily functions related to athletic performance. For example, it helps to regulate blood pressure, muscle contraction, and the concentrations of glucose and calcium in the body. Scientists believe that nitric oxide helps to improve athletic performance because it dilates the blood vessels, which increases blood flow through the muscles and boosts the supply of oxygen. Exactly how nitrate does all this is still a mystery, although recent studies have shown that it has nothing to do with the way in which mitochondria boost energy production in the muscles. The exact mechanism is yet to be revealed, but it appears to allow amateur athletes to use less oxygen for the same amount of effort. Nitrate is found not only in beetroot but also in leafy vegetables like lettuce, endive, purslane, rocket, and spinach. The ADI (acceptable daily intake) for nitrate is between 0 and 3.7 mg per kilo (0 and 1.7 mg per pound) of body weight per day. This is the dosage that you can consume each day without causing any harm to your health. It translates into 259 mg per day for an adult weighing seventy kilograms (155 lbs). If you eat more than two hundred grams of one of the vegetables listed above in a single day, your nitrate level will exceed the ADI.

In 2007, Filip Larsen at the Karolinska Institute in Sweden was the first to demonstrate that endurance athletes require less oxygen when they add more nitrate to their diet. The dosage of nitrate he gave to test subjects in the three days leading up to a cycling test was the equivalent of 150–250 grams of spinach, lettuce, or beetroot. Beetroot juice and other nitrate supplements became very popular among athletes in the aftermath of his study.

Scientific research into nitrate really took off after Larsen published his results. In 2014, the human movement scientists Lex Verdijk and Luc van Loon at the University of Maastricht began investigating the alleged performance-enhancing effects of beetroot juice in professional athletes. The results to date have been mixed. Sometimes it works, sometimes it doesn't. However, the results among amateur athletes are overwhelmingly positive. One study in Maastricht involving well-trained amateur cyclists showed that test subjects who had been given 140 ml doses of beetroot juice over a period of six days cycled 1.2 percent faster during a ten-kilometer time trial than those who had been given a placebo of beetroot juice containing no nitrate. The question, of course, is just how significant an improvement of 1 percent is to amateur athletes. That said, most researchers have reported a slightly greater improvement (up to 3 percent) in studies with relatively untrained people.

The effect on very well-trained athletes is negligible, as was demonstrated when a group of professional athletes were given a one-off dosage of nitrate and asked to cycle a time trial. There was also no improvement after they had undergone a six-day course of supplements (800 mg per day). The amounts of nitrate and nitrite in the athletes' blood did rise, but they did not perform any better. The result was the same with a group of professional 1500 m runners. So why doesn't beetroot juice work for professional athletes? The belief is that beetroot juice has no effect on them because they already have a very high aerobic capacity.

There is no denying that beetroot juice can boost your performance, but in which sports and in which kinds of athletes is still unknown. The optimal dosage also has yet to be established. So what happens if you exceed the daily dosage currently regarded as safe? In scientific research, the dosage used is usually far higher than the ADI. Half a liter of beetroot juice can contain up to 300 mg of nitrate, depending on the brand. That's on top of the nitrate you get from your vegetables. Beetroot juice is also available in concentrated form, which can push your nitrate levels way above your RDI (reference daily intake) very quickly. The Beet-it company produces small 70 ml bottles of concentrated beetroot juice specifically for athletes and these contain a whopping 400 mg of nitrate, which is more than you will find in a 500 ml bottle of normal beetroot juice. Is that a good idea? Exceeding your

RDI once or twice won't do you any harm. But what happens when you take high doses continuously for several weeks? Toxicologist Theo de Kok at the University of Maastricht has expressed his concerns in various media. In an interview with the Dutch newspaper *de Volkskrant* he said that the long-term risk of overdosing on nitrate has never been studied. By this he means the risk of cancer. Nitrate is known to cause carcinogenic nitrosamines in the stomach and intestines. In his opinion, drinking excessive amounts of beetroot juice is like "playing Russian roulette with your health." The Netherlands Nutrition Center advises people not to use nitrate-rich supplements like beetroot juice on a daily basis.

One more thing you should consider if you are a fan of beetroot juice: if you suffer from low blood pressure, you should avoid it. Red beetroot juice, arugula, and spinach are all known to reduce blood pressure. If your blood pressure is already very low, drinking beetroot juice can even cause you to faint.

MORE DOES NOT MEAN BETTER

The list of supplements does not stop at beetroot juice. What about magnesium, iron, sodium bicarbonate, and vitamin pills? What can we say about these? With regard to iron and magnesium, sports dietician Jenny Hofstede says that there is no evidence to suggest that it improves performance as long as the athlete is not suffering from a deficiency. Some runners take magnesium in the belief that it helps prevent cramps. This has never been proven. "Sports-related muscle cramp is often the result of a lack of fluids or suddenly deciding to run much longer distances," according to Hofstede.

Iron deficiency and anemia are common among athletes. However, taking extra iron is necessary only if you still lack the mineral despite making iron-enriching adjustments to your diet.

It's a bit more complicated when it comes to sodium bicarbonate (also known as baking soda). Bicarbonate works as a buffer against lactic acid in the muscles when taking intensive and explosive physical exercise that lasts no more than a few minutes. Imagine you are running a one-mile race. Your leg muscles acidify when the lactic acid produced is converted into lactate and a H+ ion, with the latter being the main culprit. Acidification causes the enzymes in the muscles to work less efficiently, which has a dramatic effect on performance. Bicarbonate neutralizes everything by removing the H+ ions. Athletes also like to use the supplement beta-alanine because of its alleged buffering properties. Supplements that buffer lactic acid are most effective in the case of explosive sports, like sprinting. They are of little use to the average marathon runner, who rarely suffers from acidification.

What about vitamin supplements? I know people who take multivitamins every day because they think "What's the harm?" Trainers sometimes advise their athletes to take extra vitamins as a preventative measure. It is true that you need more vitamins and minerals when you train hard, says Hofstede, but you should be able to get all you require out of a proper, balanced diet. There is also no point in taking vitamins when your vitamin status is already good; more pills will not improve your performance. In fact, taking too many vitamins can be harmful. Experts advise against exceeding the RDI, which is exactly what happens when you take certain vitamin pills. Take a quick look at the shelves in any drugstore and you will see that some pills contain 1,000 mg of vitamin C. The equivalent of eating three melons! In the Netherlands, the RDI for vitamin C for adults is 75 mg per day. A single 1,000 mg pill, therefore, will give you over ten times more vitamin C than you actually need. Most people don't give this a moment's thought. Nor do they realize that too much vitamin C can increase the risk of kidney stones, or that an overdose of vitamin B6 can lead to pain and sensation disorders. In 2010, the Dutch speed skater Sven Kramer suffered nerve damage to his right leg—the result, according to medical experts, of swallowing sixteen times the RDI of vitamin B6. Kramer had been taking the pills on the advice of an orthomolecular doctor. "An overdose of B6 can cause muscle tension and cramp," according to Hofstede. "It might sound obvious, but I would only recommend taking vitamin pills when there is a real reason to do so, like when a blood test shows up a deficiency."

Two conclusions to close out this chapter: when it comes to supplements, more does not mean better. And, more importantly, for the dedicated runner a well-thought-out and healthy basic diet does more good than all the pills and powders in the world.

TIP

Sport-specific nutrition can make a difference for runners who train between three and seven hours a week. Make sure you have: enough carbohydrates during the day, a high-carb meal before training, something to quench your thirst during training, and a recovery meal containing proteins and carbohydrates when you are done.

7

A SPRINT TO THE JOHN

Pain is inevitable. Suffering is optional.
—Haruki Murakami, *What I Talk about When I Talk about Running*

In the world of running, Paula Radcliffe is famous first and foremost because of her impressive athletic performances. She has run world records for the marathon and is a six-time long-distance world champion. In 2015, this living legend called time on her running career at the age of 41. Despite the fact that she has often achieved the seemingly impossible, she fears being remembered most for a second and annoyingly persistent reason, as she explained in a newspaper interview. Radcliffe was referring to the "poop incident" during the London marathon in 2005 when she suffered stomach cramps that got worse and worse as the race went on. Without breaking stride, she searched in vain for a portable toilet along the route. Eventually the cramps became so bad that she had no choice but to stop at the side of the road and relieve herself in full view of the watching public.

She did what she had to do. And with no regrets. She won the race.

Paula Radcliffe's tale will sound uncomfortably familiar to many runners. I once suffered stomach cramps that got steadily worse while taking part in a run in Boston. I even had to stop and walk part of the route because I could barely stand up straight. When the cramps became a little less severe I tried running again in the hope that I might at least be able to make it to my own house. That quickly proved impossible. As I was crossing a bridge I spotted a university building. Bent almost double, I dragged myself up the steps, past the stone pillars and under the magnificent dome, where I finally found the sign I was looking for: "Restroom." The stalls were open at the top and bottom, but I couldn't have cared less. I locked myself in for what must have been a good fifteen minutes. Luckily, none of the students seemed in any way put out by the appearance of a stray runner in their midst. But where did those cramps come from all of a sudden? Why do you often feel the need to poop when you're out for a run? And is there anything you can do to ward off stomach problems?

TROUBLESOME TUMMY

The list is long: nausea, indigestion, flatulence, vomiting, stomach cramps, feeling bloated, a stitch in your side. But worst of all is the sudden, overwhelming urge to defecate, like what happened to Paula Radcliffe. The pain gets worse with each step; all you can think of is finding a restroom as quickly as possible. Long-distance runners often have to endure these symptoms during intensive training sessions and races. The stomach-liver-bowel specialist Rinze ter Steege did his doctoral thesis on exercise-related stomach and bowel problems among recreational runners in 2012 at the UMC in Groningen. In 2006, he asked 1,300 participants at the Enschede Marathon to fill in a survey. Eleven percent said they suffered "severe" stomach and bowel problems during the race and listed several different symptoms. Almost half of the runners indicated they had suffered at least one kind of problem, including stomach pains, diarrhea, and cramps. The most common complaint was side stich. The day after the race, 3 percent of the runners were still suffering problems.

Not enough oxygen in the digestive system is a common cause of stomach complaints. In the case of intensive exercise, most of your blood travels to the muscles and skin. Less blood flows to the gastrointestinal tract, which results in a shortage of oxygen. Without enough oxygen, it's impossible to digest your food, or even make the gastric juices required to do so. In one of ter Steege's studies, a lack of oxygen turned out to be the guilty party among twelve patients treated for stomach problems after taking exercise.

When I call him on the phone, ter Steege explains the phenomenon in more detail. "What happens is that the gastrointestinal tract becomes less 'agile,' so that your stomach empties more slowly during an intensive training session. In other words, if you pour a large amount of water into your stomach while running at high speed, it will just sit there. And that's what gives your stomach that bloated, sloshy feeling." The sphincter at the end of the esophagus also appears to remain open more often when you are running, which can cause you to belch. In addition, the capacity of the small intestine to absorb sugars is also affected. Ninety-five percent of the food we eat is absorbed by the small intestine, and you can suffer problems if the amount of gels you consume exceeds the capacity of the small intestine to absorb carbohydrates. In that case, the small intestine transfers the surplus to the large intestine, where the carbohydrates are fermented by millions of bacteria. The result? Stomachache and diarrhea. The large intestine is forced to work harder, and that makes you want to poop while running. "That's why exercise is said to be a good remedy for constipation," says ter Steege. Nerves and stress can also strengthen the urge to run to the nearest restroom.

A lack of oxygen can have a disastrous effect on the large intestine's mucous membrane, and marathon runners sometimes notice blood in their stool. Fortunately, the underlying layers and muscles are normally not affected, so you don't end up with a perforated bowel. But is this blood loss the reason why endurance athletes are more prone to iron deficiency and anemia? "No," insists ter Steege. The amount of blood lost is never more than a few milliliters. "Anemia in endurance athletes may be the result of the breakdown of red blood cells caused by exercise," at least according to his own theory. It is not altogether clear why athletes are more prone to anemia, but it is probably because of a number of factors, such as iron lost through perspiration and, in the case of women runners, heavy menstrual periods. Pills like ibuprofen and diclofenac, which runners sometimes take to keep muscle and joint pain at bay, can also cause bleeding in the gastrointestinal tract.

Another aspect of running can also contribute to stomach problems. "All that bouncing around really shakes up your insides," says ter Steege. Other athletes are affected by the same problem too. Jockeys, for example, often suffer gastro-intestinal complaints. "However, a jockey's heart does not beat as fast as a runner's. Most of the symptoms seems to be the result of a high heart rate combined with too much shake, rattle and roll."

We are not all equally sensitive to this kind of trouble. Ter Steege's own research has shown that women and runners under twenty years of age suffer the most, as well as people who eat and drink while taking exercise, especially if they're not used to doing so. 'Young people have a higher level of sympathetic nerve activity, which reduces blood flow to the gastrointestinal tract, and their faster heartbeat allows them to go 'deeper'—two reasons why they suffer more quickly from a lack of oxygen.' It is also possible that today's older generation of runners is made up of those who had no prob-lems in their younger years—a 'survival of the fittest' sce-nario. Why women are more prone to these complaints than men is still a mystery. 'What we do know is that women suffer more from stomach problems like irritable bowel syn-drome.' Fortunately, most gastrointestinal problems clear up pretty quickly without causing any lasting damage. But that does not make them any less irritating, of course. So is there any way you can keep a troublesome tummy in check?

TRAINING YOUR GUT

You can train your heart, lungs, and muscles to cope bet-ter with long distances and demanding races. Most scien-tists believe that you can also train your stomach and gut to take on large amounts of water and nutrients while running.

When your digestive system becomes used to eating and drinking while running, your stomach can empty itself faster and the small intestine will absorb carbohydrates more efficiently. The idea is to eat lots of carbohydrates before training and to make this part of your weekly routine, and to get used to eating carbohydrates while exercising too. A well-trained gut will reduce stomachache and even boost performance. Two birds with one stone, in other words. Okay, this all sounds very convincing, but where's the proof?

The evidence is mostly anecdotal, but studies carried out on animals show that the gut can be trained. Ter Steege believes that people can learn how to digest food better through repeated practice, even though there is still no hard scientific evidence.

The first time I ate a gel during a run, I did so without heeding the advice of my trainer to try it in training first. Thankfully, I didn't suffer any problems, but it is advisable to get used to taking them, according to Ter Steege. 'During strenuous exercise, your stomach empties itself slower, meaning that the gel could come back up unexpectedly. That said, there is evidence to suggest that well-trained athletes have become better at digesting food more quickly.'

Preventing stomach complaints is primarily a matter of heeding good advice: not too much, not too heavy, and not too soon before exercise. The evening before a run, it is best to avoid foods that cause gas, such as beans and cabbage. 'Eat your last meal at least two hours before and don't eat any fats, proteins, or fibers during a run. Fats slow down the process of emptying the stomach and fibers draw water into the intestines, which can feel uncomfortable. Fibers also make you want to poop.' Hypertonic drinks, which contain high concentrations of sugars and salts, should also be avoided during races, as they are absorbed too slowly into the blood. 'Hypertonic drinks delay gastric emptying, meaning that the fluid sloshes around in the stomach for longer. I suffered from this once myself and, believe me, you don't run any faster on a stomach that won't empty,' says Ter Steege. Also, if you are prone to stomachache, it is best to avoid caffeine before and during a run.

The only other useful advice is: go to the toilet before you set off running so that you don't end up feeling the urge 'to go'. Easier said than done, of course. I always find myself sitting unsuccessfully on the john before heading off to a race, especially when I'm away for the weekend and don't have the comfort of my own throne at home. There are usually plenty of portable toilets at running events, so you will at least have a second chance before the starting gun sounds. Doing your warming-up first also helps to get the bowels moving and increases your chances of success.

CRUCIAL CREATURES

There's a lot of talk these days about the microbiome: a collective name for the four and half pounds of bacteria and other microorganisms that live on and in our bodies, the majority of which are to be found in the large intestine. The microbiome appears to play a role in everything that has a bearing on our health; from bodyweight to mood to immune system. But what have the bacteria in your intestines got to do with running? In 2017 two Spanish scientists published an article in *Journal of Sport and Health Science* that outlined the link between endurance sports and intestinal flora. Their conclusion? It is highly likely that the microbes in our intestines play a role in determining how our body reacts to physical exercise.

A vigorous run demands a lot from your body. The balance of electrolytes (minerals) becomes disturbed, the supply of glycogen is plundered, the mucous membrane of the intestines becomes damaged, as do the muscles, and inflammation and oxidative stress rear their ugly heads. Intense physical exercise causes the cells to produce large amounts of ROS (reactive oxygen species) that attack the cell—not exactly a great advertisement for the benefits of running. However, as soon as all this demolition work has been completed, the repair work kicks in, with the aim of coming back even stronger than before. And this is where our intestinal bacteria play an active role. It is a well-known fact that these little creatures have a crucial impact on our metabolism; they make it easier for us to break down carbohydrates, proteins, and fibers. A healthy microbiome is also important to the efficient functioning of the immune system and the nervous system, although it is not known exactly why or how. Many experts believe that a healthy bacteria population enables a runner to release energy quicker during a run and to recover quicker afterwards too.

Scientists working at the George Church lab at Harvard University have spent the past few years studying the microbiomes of exceptionally talented athletes and comparing their results with the bacterial composition of non-athletic people. The team has been able to identify a number of bacteria that may play a role in athletic performance. Athletes who had trained for the Boston Marathon and were tested after the race had a much higher level of a certain type of bacteria in their poop, one that is known to break down lactic acid. The scientists are now planning to feed the bacteria to mice to see if they produce less lactate and become less fatigued from running. Their dream is to develop probiotic supplements for athletes that will help them turn nutrients into energy more efficiently.

Are some athletes just lucky to have certain microbes living in their intestines? Or has their athletic career been a determining factor? The latter is not impossible. The balance between the numbers and different types of microbes that live in the digestive system can be altered in a positive or negative way by nutrition, medication, and stress. And also by taking part in sports. A recent study conducted on both mice and humans at the University of Illinois revealed that physical exercise can alter the bacterial composition in the intestines. The microbiome of an obese individual also reacts differently to exercise than the bacterial population of a slim person. Ultimately, it seems to work both ways: running can alter the composition of the microbiome, and a more diverse microbiome probably helps to improve athletic performance.

There is also another way in which your intestines profit in the long term from running. There is plenty of evidence to suggest that physical activity reduces the risk of bowel cancer. The beneficial effect of exercise applies to every kind of physical activity, including recreational sports like jogging, cycling, swimming, and walking, as well as labor-related activities, such as lifting and digging. The theory is that the food consumed by active people spends less time in the intestines. This means that the mucous membrane of the intestine is exposed to food for shorter periods of time, which may eliminate the risk and possible consequences of infection.

DRINK BEFORE YOU GET THIRSTY

These words of advice may sound familiar: "Drink before you get thirsty, otherwise it'll be too late." "Too late for what?" is usually my first reaction whenever I hear this. I rarely drink when I am running. During a 5 or 10K race I usually ignore the water stations, unless the weather is unbearably warm. Even during a half marathon I rarely drink anything. If I'm not thirsty, I prefer not to disrupt the flow. My running companions think I'm crazy. After all, dehydration is bad for your time and your health. But how bad, really?

The strange thing is that in the 1970s marathon runners were advised against drinking too much because it would only slow them down. Today, however, we are bombarded with the advice to keep drinking. In an article in the *British Medical Journal* the journalist and academic Deborah Cohen criticized the sports drink industry for overstating the importance of drinking. In a piece entitled "The Truth about Sports Drinks," she describes how science and industry have worked together to promote "the science of hydration." And not without success. Today we all believe that sports drinks are essential. But where's the proof? The answer: there is none.

According to Cohen, all the focus on hydration can be traced back to the rise of jogging. Nutritional supplement companies were quick to spot an opportunity. Dehydration had never been a hot topic in science until then. Today, the sales figures for Gatorade, the first drink on the market that claimed to combat dehydration, continue to rise, thanks in no small part to very clever marketing littered with scientific jargon. It would almost make you think that sports drinks are the best-researched food product in the world, suggests Cohen. So what has been the main success of the Gatorade Sport Science Institute? The propagation of the idea that the human body does not know itself when it is thirsty, despite the fact that we have an excellent mechanism for detecting and reacting to dehydration: thirst.

DEMYSTIFYING DEHYDRATION

The 1990s were good years for the scaremongers, who availed themselves of every opportunity to inform athletes of "the dangers" of dehydration. They also proclaimed that water was not the best fluid for rehydration. Sports drinks were the answer. "In fact, dehydration is not life-threatening," says Arthur Siegel, Professor of Medicine at Harvard University and an advisor to the Boston Marathon, in the article written by Cohen. "Dehydration is a normal biological response to exercise." Furthermore, there are always enough fluids available during a race, should a runner become thirsty. "It's not like they are stranded in the desert with no access to water." He also points out that drinking too much is more dangerous than becoming a little dehydrated.

The evidence shows that no one has ever died of dehydration as a result of running a marathon, but sixteen fatalities and 1,600 patients have been recorded among professional marathon runners as a result of not having enough sodium in their blood. This is what happens when athletes engaged in prolonged physical activity drink more than their kidneys can tolerate. The sodium in the blood becomes diluted, which is not good news for the brain: confusion, convulsions, and loss of consciousness often follow. On rare occasions, a lack of sodium in the blood can even be life threatening. Hyponatremia, to give it its proper name, has been identified by scientists as one of the most common medical complaints associated with long-distance running. Sports drink manufacturers claim that water is the source of the problem, but a review of the literature from 2007 shows that hypotonic drinks rich in electrolytes (salts) do not help to prevent hyponatremia. So it is not the composition of the fluid itself but the excessive amount that is most likely the problem. A study of marathon runners in Boston revealed no link between a fluid's composition and hyponatremia. The runners most at risk were those who weighed more at the finish than they did at the start of the race because they drank too much.

The human body contains sensors that continuously monitor the volume of blood and the concentration of salts and minerals. If the volume falls or the concentration of electrolytes rises, the sensors send a signal to the brain, where the incoming message is translated into a conscious sensation we experience as thirst. You begin to feel the sensation of thirst when you lose 8 to 10 percent of your blood plasma, which equates to 0.5 percent of your body weight. In 2011, exercise physiologist Eric Goulet at the University of Sherbrooke in Canada carried out a meta-analysis of the data taken from cyclists in a time trial. He found that they performed best when they drank because they were thirsty, not when they drank automatically from their bottle every ten minutes. Today the American College of Sports Medicine

(ACMS) recommends that athletes allow themselves to be guided by their own thirst. Back in 1996, however, the ACSM's advice to athletes was to drink as much as the body can tolerate. This was not long after they had accepted a donation of $250,000 from Gatorade.

Experts now say that you shouldn't lose more than 2 percent of your body weight, though there is still little evidence to support this claim. In his study of the literature, Goulet found that dehydration that resulted in a loss of up to 4 percent of body weight did not adversely affect the performance of cyclists. This means that someone weighing 165 pounds can lose up to six pints of fluid through sweating while cycling without their performance suffering as a result.

When the weather is hot I can get thirsty very quickly during a run. When the temperature is low, however, I can run for one and a half hours without feeling thirsty and having something to drink. This appears to be perfectly normal. You can assume that if you don't feel thirsty, the level of dehydration is negligible. Of course, it is important that your body does not dehydrate too much. Depending on body weight and temperature, when you are running you can sweat out somewhere between one and four pints every hour. If you are running a long-distance race like a marathon, drinking enough is very important. It stops you from overheating and keeps the amount of blood circulating in your body at a normal level. Hard-working muscles need plenty of blood when you are exercising, as do your stomach and intestines. In the 1980s, Professor Fred Brouns at Maastricht University discovered that marathon runners often suffer stomach problems when they lose more than 4 percent of their body weight.

Rinze ter Steege also believes that losing fluids in the short term is not a problem. "There are no water stations on the annual four-mile run in Groningen, for example. It's not a long enough distance for you to suffer the kind of dehydration that can affect performance." It's a different story for very long races, however. "If you don't drink anything for a long period, you increase the risk of cramp." So running for one and a half hours without drinking is not a problem, as long as it's not too hot. If I want to run for longer, however, I will have to change my ways and get used to pausing for a drink.

RINSING VERSUS DRINKING

The sports drinks industry may be guilty of overstating the importance of its products, but we also have to acknowledge that if it wasn't for the cooperation between industry and science, research into sports nutrition would have struggled to get off the ground. And if that was the case, we would have been denied this humorous little piece of knowledge: you don't always have to swallow sports drinks to benefit from the carbohydrates they contain. Rinsing the drink around in your mouth for five to ten seconds and then spitting it out is often a better way of boosting your performance.

In 2004, the first study was published in which cyclists were shown to complete a 10K circuit faster after rinsing with carbohydrates. The same applied to runners who rinsed with sugar compared with those given a placebo. How is this possible? Don't carbohydrates only supply you with energy after they have been absorbed by the intestines? Scientists believe that we have sensors in our mouth that are linked to the reward system in our brain. When these sensors react to sugar, a wave of pleasure is triggered in the brain. This is probably what makes everything feel a little less uncomfortable when you are running. Rinsing appears to have the most effect when you are engaged in intensive exercise for more than an hour.

Rinsing may also be more beneficial than swallowing if you suffer frequently from stomach problems while exercising. One important footnote: to keep your energy levels up during a (half) marathon, you simply have to swallow something. In the case of a long race, rinsing has the most impact in the last three miles, when your glycogen reserves are exhausted and there's not much point in eating anything because the resulting energy takes too long to travel to the muscles. Rinsing and spitting out can give you that final mental energy boost when you need it the most.

ALCOHOL

A word or two about a different kind of drinking. It's long been a part of the culture of many sports to have a drink to celebrate a victory. My trainer is dead set against this practice and advises his pupils not to treat themselves to a beer after completing a race. So is it really a bad idea to consume alcohol after you exercise?

Sports science has not yet delved deeply into the subject of alcohol. Studies involving animals suggest that alcohol disrupts certain molecular processes that are important to adaptation after taking exercise. There has been scant research involving humans, but in one study, carried out by the RMIT University in Australia, eight men were given an amount of alcohol equal to drinking twelve beers after an intensive training session. The results showed that the alcohol delayed protein production in their muscles. Even when they were given a protein shake to drink as well, the alcohol still impeded muscle recovery. Twelve glasses of beer is an awful lot, of course, and I've never seen a runner drink that much after a race. But what about one or two beers? According to Assistant Professor Floris Wardenaar at Arizona State University, even a small amount of alcohol can affect recovery. In a study conducted in 2005 on ten subjects, he found that the amount of lactate in the blood after an intensive bout of training fell slower after drinking alcohol. We met up on Skype to discuss the relationship between recovery and alcohol. "Lactate is usually cleared very quickly," he tells me. "Active recovery facilitates that process. For instance, it is a good idea to continue with some gentle jogging after finishing a race in order to clear the lactate you have built up. Clearance takes longer if you sit down immediately after the finish." Drinking a few glasses of beer means that the lactate remains in your body for longer.

The test subjects in his study had the same amount of alcohol in their breath from drinking two glasses of vodka after training as they would have had from drinking three glasses without taking any exercise. When we drink after exercise, therefore, the alcohol hits us much harder. "The enzymes in the liver that break down lactate are also responsible for doing the same with alcohol," explains Wardenaar. After running a race, the liver starts clearing up all the lactate. If you pour a beer or two into the mix, it takes longer to clear up both the alcohol and the lactate.

A recent study that Wardenaar conducted with Marco Mensink and students at the University of Wageningen showed, not altogether surprisingly, that beer disturbs the body's balance of fluids. Alcohol is a diuretic—it makes you pee more—while what you really should be doing is replenishing your stock of fluids. Drinking an isotonic sports drink like Gatorade or Lucozade Sport—which contain the same concentrations of sugars and salts as the body and can therefore be quickly absorbed—is the best way to replace lost fluids. The next best are water and beer with a very low alcohol percentage. Normal beer is the worst option.

An important question here is just how damaging delayed recovery is for the body. Wardenaar believes that it is mostly a problem when you are competing in a series of races. "If you run several days in succession, you have very little time to restore your energy and fluids and to recover before you have to run again." In that case, drinking alcohol can have a negative effect. "We also know that your priorities shift considerably when you drink a lot of alcohol after a race. It causes you to behave differently and to pay less attention to eating and drinking the right things. Whether or not you've been taking care of yourself becomes more and more obvious as the days and races go by," says Wardenaar. Optimal recovery is less crucial when you have two weeks to recover after running a marathon and don't need to replenish your sugar reserves with any great haste. In that case, an alcoholic drink after the race should be no problem. The literature also suggests that alcohol does not disturb the replenishing of glycogen reserves. Nevertheless, Wardenaar recommends consuming plenty of fluids, electrolytes, protein, and carbohydrates first, after running a (half) marathon, before you order your first beer.

TIP

Practice eating concentrated carbohydrates (gels) and other foods and avoid fiber, fat, protein, and gas-producing foods before a race as much as possible. Take a trip to the restroom before a race and do a warm-up to get your bowels moving if need be. Finally: while running, keep what you eat and drink to a minimum. Even when you get better at taking water and food on board during exercise, remember that there is a limit to what the body can process per hour.

8

RUNNING FOR YOUR LIFE

There is more to running than simply "not dying."
—Koen Breedveld, former director of Mulier Institute

The Hadza people of north-central Tanzania are modern-day hunter-gatherers. Their nomadic lifestyle differs in many ways from ours, with the biggest difference being in daily activities. They engage in an average of 135 minutes of moderate to very intensive exercise each day, which is four times more than what most health authorities in the West recommend. In the Netherlands, for example, people are advised to take 150 minutes of exercise a week, spread over several days. The Hadza remain active all their lives, with both the elderly and the young used to covering very long distances on foot. Scientists who visited the tribe discovered that the Hadza have an extremely low risk of heart disease. Of course, a lower risk of heart disease does not come from exercise alone but depends on other factors as well, like diet, smoking, alcohol, and stress. However, there is definitely a strong relationship between getting enough exercise and having a healthy heart.

In the Netherlands, 1.5 million people supplement their weekly dose of exercise by going for a run at least once a week. Bravo! International studies have shown that people who run live longer than those who don't. In fact, it even appears that no other form of physical exercise is as beneficial in terms of longevity as going for a run. That said, no two runners are the same, and there are large differences between the number of miles individuals run. While some of us like to go for a gentle jog twice a week, others train hard four or five times a week with the aim of running faster and farther. At the very end of the spectrum are the dedicated mile crunchers who like to run ultramarathons; a select group for whom running sixty miles a week is nothing special. So is all this running good for your health? Or is there a limit to how far and how often you should run? Is there a point beyond which intensive exercise is actually bad for you? Let's see what the science has to say.

FROM ZERO TO HERO

Countless studies have shown that those of us who get regular exercise live longer than those who sit all day. I like to think I have an active lifestyle, but, to my great disappointment, the facts appear to tell a different story. I start my working days by sitting down for breakfast. I then take no more than three steps and sit down again at my computer, where I often spend up to eight hours a day staring at the screen. Okay, I do set an alarm to remind me to stand up and stretch every forty-five minutes, but that's about as strenuous as it gets. If I weren't an avid runner and used to spending a lot of time in the gym, my body would hardly move at all, apart from the two minutes it takes me to walk to the nearest supermarket. Without the sporting activities I engage in, I would probably collapse on the couch every evening to watch TV or read a book. Some people think that exercising five times a week is way too much, but I see it differently, especially when I think of the amount of time I spend sitting down every day. I regard the exercise I take as nothing short of essential.

Experts say that getting enough exercise is the best strategy for preventing and dealing with a whole range of health problems. "If we had a drug in our therapeutic armamentarium that conferred all the benefits of regular exercise, it would, arguably, be the single best treatment for preventing disease and improving overall health and life expectancy," according to a team of Danish and American scientists writing in the journal *Heart*. The link between exercise and health is beyond doubt. Physical effort reduces the risk of stroke, heart disease, breast and bowel cancer, and type 2 diabetes—five common chronic illnesses. The health benefits increase greatly when the total amount of exercise exceeds the recommended minimum. So whether we want to develop muscle tissue, build stronger bones, or combat stress, one thing is for sure: exercise helps. And the more you get the better. The accepted guideline is 150 minutes of exercise per week, spread over several days. This involves taking moderately intensive exercise, like cycling or walking briskly. We are also advised to add extra muscle- and bone-strengthening activities at least two times a week.

Only about 23 percent of all adults in the United States take the recommended amount of exercise each week. Encouraging people to add even the bare minimum of exercise to their routine is still a very difficult task. Professor Daniel Lieberman at Harvard University—who believes, as already mentioned in chapter 1, that humans have evolved to run long distances—has a plausible explanation for why it is so difficult to get people moving. He says that although humans may have evolved to run long distances, we are equally good at being inactive. Natural selection has ensured that we do all we can to avoid unnecessary effort. We prefer to be lazy than tired, despite the benefits of exercise for our health. He refers to this phenomenon as the *exercise paradox*. Inactivity actually makes sense in terms of evolution. Our ancestors had a limited supply of food available to them and they reserved their energy for essential activities like hunting and gathering. In between, they rested as much as possible and did not waste any precious energy on unnecessary tasks. That instinct is still ingrained in us and has even been reinforced through the advent of technology. We will run to catch our train if necessary, but when we arrive at our destination we prefer to take the escalator rather than use the stairs.

What we should be doing, however, is grabbing every possible opportunity to move our butts. Getting more—and more intensive—exercise than the recommended 150 minutes a week results in lots of extra health benefits. Fifteen minutes of moderately intensive exercise a day is enough to considerably reduce the risk of heart disease and premature mortality. And the more exercise you get, the greater the cardiovascular benefits.

Thijs Eijsvogels, an exercise physiologist at Radboud University Medical Center in the Netherlands, spends much of his time studying the relationship between physical activity and health. One of the main questions on my mind while writing this book related to the role that sport plays in the story. Is it even necessary to go jogging when you already get enough exercise just by cycling to the store and gardening? Eijsvogels's answer leaves little room for doubt: "The more you move, the less likely you are to suffer health problems. At least up to a limit of 100 minutes per day. We don't see any causal link between exercise and extra health benefits above that limit." Intensive effort and buckets of sweat are not always necessary, he explains. "The main health benefits come when you switch from sitting around all day to engaging in daily physical activity, for instance by cycling to work or going for a walk during your lunch break." Actively participating in sports can also improve your health and lengthen your life span. "The fitter you are, the lower your risk of illness," according to Eijsvogels, but when you are fit the health benefits are not as dramatic as those you achieve when you first start exercising and go from zero to hero.

USE IT OR LOSE IT

There is mounting evidence that sitting increases the risk of disease, in particular diabetes. It doesn't matter how hard you train if you sit too much, as they both influence your health in different ways. But you are in real trouble when you spend a lot of time on your rear end *and* get very little exercise. Even just getting up from your chair more often helps. This is why people are being encouraged to sit less nowadays.

But why is sitting so bad for us? I am well aware that inactivity is bad news for your muscles. You can lose up to four pounds of muscle mass when you are bedridden for five days. But why do the active ones among us live longer than the sitters? A good question for the human movement scientist Hans Savelberg at the University of Maastricht. Sitting all day is bad for our sensitivity to insulin, according to studies carried out at his lab. Insulin is the hormone that pumps the sugar from our blood into our cells. Low insulin sensitivity is an early sign of type 2 diabetes. "When you don't use the cell receptors to which insulin binds itself, they are cleared away," explains Savelberg. "You must take good care of the system if you want to keep your insulin sensitivity in check. Getting up from your chair regularly helps to keep everything alert. All biological systems are subject to the same maxim: use it or lose it." Our body needs to be stimulated regularly in order to keep systems like blood circulation and metabolism in good working order. The stimulant does not have to be sports-related, either. A brisk walk or climbing the stairs is often enough, but you can also improve the condition of the body's systems by providing an even stronger stimulant, for example by lifting weights.

THE UPPER LIMIT

We have established that the slogan "More exercise is better" certainly applies when it comes to physical activities like cycling, gardening, cleaning, and walking. But does it also apply to running? Is running faster and farther better than going for a gentle jog? If exercise really is a kind of medicine, there must be a relationship between dosage and effect. A low dosage of a medicine has little or no effect, while an overdose can be very dangerous. Somewhere between these two extremes lies an amount that gives you all of the gain and none of the pain. If you live a relatively active but sport-free life, you are unlikely to end up overdosing on exercise. The story is a little different, however, for endurance athletes who participate regularly in triathlons and marathons. I run three times a week, covering around twenty miles in total. But that number goes up when I am training for a long race like a half marathon. The question is: Does this fall within the therapeutic window for effective dosage? Where is the line between healthy and unhealthy?

Too much exercise is probably not good for your life span. An international study in 2015, in which scientists examined the relationship between the amount of exercise and the resulting health benefits, revealed what appeared to be an upper limit. They examined the data from six different studies involving a total of 600,000 people in Europe and the United States. The health benefits reached a maximum when the exercise taken was three to five times greater than the minimum recommended level and stagnated when the level of exercise was higher. Ten times more exercise than the recommended amount—the equivalent of twenty-five hours of exercise per week—actually resulted in a dramatic fall in health benefits compared with normal levels. Nevertheless, the risk of premature death in these extreme cases was still lower than for those who spent most of their time sitting down.

Another study, the Aerobics Center Longitudinal Study, which looked specifically at running, concluded that the upper limit was twenty miles a week. Running farther did not reduce the risk of premature death anymore. A "lower dose" of running—below the twenty-mile limit—appeared to be more beneficial in terms of life span. While it is difficult to specify an exact upper limit, it is probably only a matter of time before someone comes up with the magic number.

The benefits of jogging have been championed in many different studies. Those that have followed large groups of test subjects for a number of years reveal that active people are 30 percent less likely to die prematurely than their inactive counterparts. The overall conclusion is that people who run live three years longer than those who don't. Running is strongly linked with a lower risk of premature death as a result of heart disease or other ailments. What the studies do not provide, however, is definitive proof that *running* is the reason behind that longevity. The only thing long-term observational studies have demonstrated is that people who run are also people who live longer. But maybe healthy people only run more because they feel fit, which would mean that running is not the reason behind their good health but a consequence of it. However, most scientists believe we have enough proof that running can prevent chronic diseases and premature mortality. The correlation stands up to scrutiny when you remove other factors that contribute to good health and longevity from the equation. For example, both smokers and nonsmokers live longer when they run. It also helps that there is a ready-made theoretical model that explains why jogging is so healthy. Researchers have established, for example, that running lowers your blood pressure, your resting heart rate, and your fat percentage. It also improves your sensitivity to insulin and builds up muscle mass, bone density, "good" cholesterol, and gray matter in certain parts of the brain. All of which helps to lower the risk of obesity, heart disease, stroke, cancer, and diabetes, with the end result being a longer and healthier life.

ULTRAMARATHON VERSUS COUCH POTATO

To find out more about the ups and downs of life as a fanatical runner, I decided to visit Frank Backx, sports physician and Professor of Clinical Sports Medicine at the University Medical Center Utrecht (UMCU). In 2017, the Dutch public health authorities asked him to help draw up new guidelines for exercise based on the most recent scientific knowledge. Backx runs twice a week himself, either on his own or with a running partner, so I figured that a professor of sports medicine who regularly pulls on a pair of running shoes would be a reliable source of information. I asked him if twice a week was the perfect running schedule. "It's a bit more complicated than that," was his reply. "We recommend exercising two or three times a week, spread over the entire week so that you have enough time to recover. People who have quicker recovery times often expand this gradually to four or five times a week. There are also those who get no exercise during the week but who go running on Saturday and Sunday; so-called 'weekend warriors.'" This is not very clever, says Backx, because you can never recover enough from one day to the next. However, recent studies have shown that being a weekend warrior is just as effective at staving off heart disease as training over the course of a full week. While the effect in terms of injuries has never been examined, it seems pretty logical to me that running two days in succession involves a higher risk of injury than running with rest days in between, certainly for beginners who are not yet used to the training load. Weekend warrior or not, one hour of exercise at 70 percent of your maximum capacity, with or without pausing, is the kind of effort that delivers the greatest health benefits. A level of effort, in other words, that doesn't leave you gasping for air and unable to speak. "Having a running partner can stop you from pushing yourself too far," adds Backx.

In terms of the health benefits, moderate running is probably better than extreme running, though the evidence is still lacking. Moderation is not a new concept; you can also apply it to sleeping habits, alcohol, and calories. Equally, you can also get more exercise than is actually good for you. Does this mean that extreme endurance athletes are as badly off as couch potatoes? According to Backx, running ultramarathons will not condemn you to an early grave, but it may diminish the benefits of exercising normally. "Sitting is still the worst thing you can do in terms of living longer. There is no dose of physical activity, no matter how strenuous or long-lasting, that carries as high a mortality risk as being inactive."

You don't need to engage in intensive physical exercise to maintain your health, as long as you are relatively active and don't sit around all day, according to Thijs Eijsvogels. Staying fit is a different matter, however. The amount of exercise that is best for your heart is not the same as what you need to build up a maximum level of stamina.

AN ATHLETE'S HEART

"It is important to draw a distinction between health and fitness," says Backx, while he searches on his laptop for the data that will illustrate exactly that. Getting fit requires a lot more effort than just taking the stairs instead of the elevator. But what does the word "fit" actually mean? It's all about making the heart a more efficient instrument, one that can do more with less effort. Backx: "A fit person's heart beats slower because it is able to pump more blood per heartbeat." That is why a professional athlete's heart rate at rest is lower than the usual 60 to 80 beats per minute. A trained heart can pump more blood per heartbeat primarily because of its greater volume. The muscle wall is also thicker. "The heart muscle of a professional cyclist who spends many hours training is stronger and thicker. We call it an 'athlete's heart,' although the term has often been the source of confusion," explains Backx. "Fifty years ago people thought that this kind of enlarged heart was a sign of disease. Eventually it became apparent that the greater volume was a beneficial result of training."

An athlete's heart is not in any way harmful. "In the case of heart muscle disease, only part of the muscle is thicker, which can lead to cardiac arrest. In an athlete's heart, the enlargement of the muscle is more or less uniform; it becomes thicker in a healthy way, the same way that skeletal muscles grow as a result of training." If you stop exercising, however, you quickly lose the benefits of an athlete's heart, just as skeletal muscles wither when you don't use them. An athlete's heart requires regular maintenance.

The increase in volume and thickness can be seen as a bonus earned as a result of the demands that an athlete places on his or her heart. However, an athlete runs a greater risk of suffering heart problems during training because they force the heart to beat faster. This can actually end up triggering a previously undetected heart condition. "You only feel the benefits of training after the fact. And therein lies the paradox. You run a greater risk of an irregular heartbeat during strenuous physical exercise than when you are carrying your shopping bags."

A SCARRED HEART

There is still much discussion in the medical world with regard to the potentially damaging effects on the heart of running long-distance races like the marathon. After running 26 miles, a runner's blood often contains higher concentrations of troponin, a protein that is released after a heart attack. The level usually drops back to normal after a day or two without any ill effects. Of course, there are also rare occasions when a runner suffers heart problems while running. It appears that the reaction of the heart to strenuous physical exercise is usually either temporary or very rare. But what if you spend your whole life running long distances? Can that lead to permanent damage? Scientists still don't know the answer to this question. The results of studies often conflict, but there have been some interesting discoveries. For example, 6 percent of all endurance athletes examined had signs of scarring on the heart. The more marathons you run, it seems, the greater the risk of scar tissue. Just like with leg muscles, when the heart muscle is required to work hard it can show minuscule signs of damage, and having to repair that damage time and time again can lead to scars, at least in theory. Exactly how harmful this can be is unknown. It depends on where the scarring occurs and the subsequent demands placed on the heart. Not every athlete suffers scarring of the heart, Backx reassures me. In fact, it is more often the exception than the rule.

In addition to all the worrying news, science has also turned up plenty of positive results. A few years ago, sports physicians at the Saarland University in Germany studied a group of thirty-three middle-aged professional athletes, including a number of ex-Olympians and Ironman contestants. They compared them with thirty-three other men of similar age who had never participated in endurance sports. The only difference that showed up on the athletes' MRI scans was their athlete's heart. There were no other differences in terms of structure or function. They concluded that it is unlikely that spending years taking part in competitive endurance sports causes heart damage in the long term.

Another theory making the rounds is that regular endurance training increases the risk of artery calcification (atherosclerosis). Studies of middle-aged endurance athletes often show up higher levels of calcification in the coronary arteries compared with nonathletic people. Not long ago the research team led by Thijs Eijsvogels found evidence that physical exercise increases the risk of calcification in the coronary arteries. Their findings were corroborated by another team in London. However, the question remains: is this such a bad thing? It turns out that there were large differences between the types of constriction in the studies. Most of the male athletes showed a build-up of calcium in the coronary artery, while inactive people showed more signs of fat accumulation. Coronary artery constriction arising from calcium is a less serious problem. Calcium is more stable than fat and less likely to cause problems in the future, says Eijsvogels. This difference in terms of constriction may explain why athletes live longer than inactive people, despite being more prone to calcification. Frank Backx is not so sure. "We cannot say for certain that this condition is sports-related. Atherosclerosis can also be genetic or the result of smoking. Sometimes we are too quick to make the connection with sport at the expense of other explanations."

There is still no convincing evidence that the heart can suffer permanent damage as a result of intensive exercise, not even in the case of ultramarathon runners. So it seems we can race away to our heart's content for the time being.

RUNNING YOURSELF INTO THE GROUND

In 1977, the US author Jim Fixx published his bestseller *The Complete Book of Running*, which went on to sell over one million copies. It was thanks in part to Fixx and his book that running, or jogging as it was more commonly known, became popular with the general public in the 1970s. In his regular media appearances, Fixx sang the praises and health benefits of exercise. Active people live longer was his message. At the age of fifty-two, however, he died suddenly of a heart attack while out running. Ironic perhaps, but this unfortunate turn of events did not undermine his creed. It turned out that heart disease ran in Fixx's family; his father had died of heart failure at age forty-three. If our jog-crazy author had not taken up running, he would probably have died much younger than he did. In fact, he had actually started running because of his heart condition. Fixx was right when he said that running was more likely to lengthen rather than shorten your lifespan.

Nevertheless, the proliferation of stories in the newspapers about the deaths of seemingly healthy young athletes can be enough to make you think twice about the health benefits of sport. The amount of media attention may be a little over the top, but in the Netherlands, for example, 150 to 200 people die suddenly each year from heart failure while engaged in their sport of choice. The most fatalities are among cyclists and footballers, followed by runners. One out of every 259,000 runners in a marathon dies as a result of a heart attack. These are very low odds when you consider the number of runners that take part in the New York Marathon each year: 50,000.

The figures for the risk of heart failure when running long distances come from a US study published in the *New England Journal of Medicine* in 2012. The research team looked at the data from almost eleven million runners who had taken part in a half or full marathon in the United States between 2000 and 2010. In those ten years, only 59 people suffered a heart attack during the race, which represents one out of every 184,000 runners. Most of the incidents were reported during full marathons and usually when the finish was within sight. Not every heart attack was fatal. More than a quarter of the victims survived, bringing the risk of death down even further to one out of every 259,000. The study also revealed which group was most at risk: male marathon runners. Fifty-one of the 59 heart attacks were suffered by males, and more often during a full marathon than during a half marathon.

The fatalities did not occur out of the blue. Some of them were the result of overheating or a lack of sodium in the blood caused by drinking too much (hyponatremia). The vast majority, however, were the result of an underlying heart problem. Roughly speaking, the runners with a cardiovascular problem fell into two categories. Around half of the victims had a congenital heart condition known as hypertrophic cardiomyopathy where the heart muscle is too thick, which can sometimes lead to fatal heart rhythm problems. Many people with hypertrophic cardiomyopathy are not aware of their congenital heart condition. Some suffer heart trouble and decide to visit their doctor, who then diagnoses the problem. Others remain completely oblivious, and in that case it takes a cardiac arrest to reveal the condition, something that can happen anytime, anywhere. Cardiologists at St. George's University in London discovered that people born with a thicker heart muscle are more likely to die when at rest, including in their sleep, than when they are exercising. The group of people with hypertrophic cardiomyopathy included a surprisingly large number of young males who had suffered cardiac arrest while participating in sports.

We already know that congenital heart disorder is the number one cause of fatalities among competitive athletes under thirty-five. It's a different story, however, for athletes over thirty-five, for whom constricted coronary arteries are almost always the biggest problem. When they become blocked, the risk of the heart suffering a lack of oxygen increases. This disrupts the heart's electrical system, which causes the chambers of the heart to quiver irregularly. Blockage as a result of calcium and cholesterol build-up can be attributed to old age, but the process can be accelerated by obesity, high blood pressure, prolonged stress, and smoking.

Peter Hollander, Emeritus Professor of Exercise Physiology at the Free University in Amsterdam, believes that we ought to stop linking heart failure with athletic performance and that the media are guilty of grossly overstating the number of fatalities during marathons. The relatively small number of sports-related deaths overshadows the health benefits that millions of people enjoy by being active. Running saves many more lives than it costs, and Hollander is adamant that exercise is better for your heart than rest.

PREVENTIVE MEDICAL CHECKUP

While the chances of you running into trouble might be small, scientists have revealed that the number of heart attacks per 100,000 male long-distance runners rose by a factor of six in the United States between 2000 and 2010. How can we explain this? We have already established that middle-aged male marathon runners face the greatest risk. The reason behind the rise in heart attacks is probably the fact that running events these days attract a lot of untrained runners, many of whom run a greater risk of suffering heart failure because of their age, weight, and lifestyle. You are only looking for trouble if you think you can start running races for health reasons without any proper training. The question then is whether adults should be tested first before they take up running or some other sport. Would that help reduce the number of fatalities?

Each year, approximately 27,900 people in the Netherlands undergo some form of preventive medical checkup, and they are not all young, super-fit athletes. Preventive measures with the aim of improving athletic performance are becoming very popular among the general public. A sports-specific preventive medical checkup starts with a long list of questions concerning your medical history, participation in sports, and susceptibility to heart disease. This is followed by a physical examination to test the nimbleness of your limbs, a lung function test, blood samples, and a test to establish your maximum training load. An ECG is also performed to measure the heart's electrical activity both at rest and while exercising.

Frank Backx believes that a preventive medical checkup can help to solve a whole range of physical problems—a bit like an annual vehicle inspection to check for faulty and rusty parts, but for your body. "Not only your heart but your whole body is screened to check for weak spots like arthritis, bad ankles, or poor balance." A sports physician can use the test to advise against certain kinds of activity and recommend the best training methods. "The test is not just a good idea for people who are worried about their heart," says Backx. It is also useful for those who are thinking of switching to a different sport or are taking up sports for the first time. "Start2Run clinics, for example, are always full of people who think it would be a good idea to take up running. In six weeks, they learn how to start out gently and then build on that. However, a quarter of them will have succumbed to injury within that period. Pushing yourself even just a little bit too far often marks the start of a downward spiral. A preventive checkup can provide you with very useful information, such as your recovery capacity."

But is the test of any value to someone like me, a relatively young runner with over ten years of running experience? A few years ago, I visited the sports physician Jessica Gal for a thorough physical examination for the purposes of an article I was writing at the time. It didn't reveal any serious problems, apart from a slight lack of mobility in my lower back. I figure there's no real need for me to go back to her any time soon. Backx does not agree. "We advise young athletes to have a checkup every five years or so. When you pass forty it's advisable to do it every two to three years, as that's the age at which your arteries begin to clog up and your tendons start to deteriorate." He cites a study carried out at the UMCU in which 20 percent of male endurance runners aged forty-five or older showed signs of congested arteries. They had no physical complaints, however. In fact, they all felt as fit as a fiddle.

It is still not clear whether congested arteries are a portent of cardiac arrhythmia and sudden cardiac death in athletes. As Eijsvogels explained to me, athletes are often prone to a more benign kind of calcification. He is not a great fan of screenings anyway. "It all sounds so promising, but I would question the effectivity of screening given the high number of erroneous diagnoses and the costs involved." Sports physicians, on the other hand, do recommend it, but they have a vested interest in performing the tests. "You could also just go to your doctor and ask them to assess your risk of heart disease, and to take a blood sample while they're at it," adds Eijsvogels.

The big question is whether preventive tests really do save lives. With the exception of a handful of sports—boxing, diving, parachuting, and cycling, for instance—in the Netherlands it is not compulsory to have yourself tested before becoming a professional athlete. In Italy, however, such a test is mandatory. The discussion made the headlines in 2017 when the Ajax soccer player Abdelhak Nouri collapsed on the field after suffering massive and ultimately permanent brain damage. It turned out that he had a heart condition, and that the club knew about it. The question is: can you spot this kind of tragedy before it happens by carrying out a preventive checkup?

Science has concluded that there is no evidence to prove that screening programs prevent sudden heart failure in young athletes. "Sometimes a test will identify a potential victim, but this still hasn't led to mandatory screening in the Netherlands," says Backx. "The evidence is just too flimsy." Not all congenital heart defects can be spotted by an ECG, meaning you may fail to identify some high-risk cases. What happens with mandatory screening is that healthy athletes are often advised not to participate in sports. An ECG can sometimes identify something that's not actually there—a so-called false positive—resulting in a false alarm. Nevertheless, and despite the lack of convincing evidence, the European Society of Cardiology has instructed sports associations that their athletes must be screened before they are allowed to take part in competitions. If you've got a dicky ticker, chances are you won't be allowed to compete. Frank Backx, who has had plenty of experience with young athletes and heart problems, does not agree. "Should you stop someone with hypertrophic cardiomyopathy or some other congenital heart condition from doing something they love? Even though you don't know when or even if they will suffer cardiac arrest? That could be in fifteen years' time or tomorrow in the supermarket. You don't help people by telling them 'you're not allowed to join in.'"

Backx touches on something very important here: enjoyment. Up to now we've been discussing everything purely in terms of health, fitness, and life span. But a large proportion of the runners out there don't run because of the potential health benefits. No one starts training for a marathon with that in mind. People run long distances for the thrill of it, for the competitive element, to tackle stress, for the social aspect, or just for the satisfaction it gives. For me, personally, it is a combination of all of the above. The mental aspects and issues like quality of life are often neglected in studies of the benefits of running. Maybe you could compare it with eating sweet stuff. I know that all that sugar is not good for me, but I really enjoy a nice slice of cake or a delicious dessert every now and then. My body might not benefit much from eating chocolate cake every week, but my mind is not going to argue. Similarly, running very long distances week in and week out might not be entirely beneficial to your health, but what if it makes you a happier person? It is also important to enjoy the things you do from day to day. Backx couldn't agree more: "If you put a lot of time into sport, you need to enjoy it too. People who take part in ultramarathons or triathlons do so because of the enjoyment it brings, and that adds quality to their lives."

TIP

If your goal in life is to remain healthy and live as long as possible, your best strategy would be to go for a gentle jog two or three times a week. There's no need to overdo it, let alone enter an ultramarathon. If you want to stay fit, however, you need to work a little harder. Intensive exercise is not optimal for your health, but the pleasure it gives is worth something too. So if you get great enjoyment out of gasping for air at the finish line after running 26 miles, then go ahead and do it. Just don't push yourself too hard, and make sure you get enough rest in between.

9

THE SECRET TO SPEED

The fun is at the back of the pack.
—a runner's slogan (Jeroen Scheerder and Koen Breedveld,
Running across Europe)

At the club where I train, there are girls who can run the 10K in less than forty minutes. No matter how hard I train, I will never match their speed. On the other hand, there are plenty of runners at the club who will never be able to match my personal best for the 10K no matter how hard they try. And even the best runners at the club are no match for East African athletes. Kenyans and Ethiopians have long dominated the world of running, and they continue to win one medal after another. The Belgian athlete Stefaan Engels has a different talent. He does not run record times for the marathon, but he does hold the record for the most marathons run in succession. Starting in February 2010, he ran 365 marathons in 365 days—a feat that is beyond even the wildest dreams of the average runner.

Not all runners are obsessed with beating their personal best or running longer distances. Some run just to stay fit, to lose weight, or to unwind after work. However, a significant chunk of runners do run with the aim of improving their performance and beating their own records. I belong to that category. I am always jealous of runners who breeze past me during a race. I want to be able to run at that speed too. Usually I am able to resign myself to the fact that some people are simply better equipped to run faster than others. But what's their secret? Generally speaking, your physique or genetic makeup should not prevent you from running, but being very fit is a different matter entirely. Not everyone is able to run for hours on end or keep up a blistering pace, but some people can. The question is: How? Do fast runners have extremely efficient muscles and hearts? Or are stamina and athletic performance shaped by other factors?

FOR THE LOVE OF SPORT

The secret to speed may lie in a runner's love for their sport. After all, not everyone is willing to lose buckets of sweat on a purely voluntary basis. Some are willing to train in all kinds of weather, while others will stay on the couch even when the sun shines. Why are we so different in this regard? Is it in our genes? My parents are not exactly sporty types; my father plays badminton once a week and my mother does yoga. They walk and cycle a lot, but they have never been as active as their children. My mother actually hates running.

Sports psychologist Charlotte Huppertz has studied the individual differences in sport participation. To do so she examined the data for sets of twins, the classic method for studying the influence of genetic makeup and environment on behavior. Huppertz discovered that differences in sport participation among young children are primarily the result of the environment in which they grow up. The children of inactive parents tend to be inactive themselves, while growing up in an active environment stimulates active behavior. Genetic makeup becomes more important and environment less so from the age of fourteen on.

Whether or not a particular talent rises to the surface depends on certain circumstances, such as access to sports clubs and physical exercise at school. With parental encouragement, you can become a sports fanatic, even if it's not in your DNA. According to Huppertz, we still don't know which genes determine our love for sport. She has not found any genetic variants that can be linked to sport participation. Our psychological reaction to sport probably involves a whole series of genes. In the end, if a run in the park feels like an enjoyable thing to do, you are more likely to do it again.

Studies on mice also provide interesting insights into our love for sport. Scientists at the Baylor College of Medicine in Houston discovered that mice that ran a lot while pregnant gave birth to mice pups who knew how to use a treadmill when they reached adulthood. After the pups had been weaned, they were separated from their mother and moved to a different cage so that they could not copy her behavior. The young mice still proved to be fanatical treadmill runners, while the pups of inactive mothers were far less enthusiastic. It appears that the behavior of a mother mouse is transferred to the fetus. This experiment cannot be extrapolated to humans, of course, but the idea that conditions in the womb can influence the development of babies is not a new one. In an article in the *New York Times* the leader of the research team, Robert Waterland, speculated that the development of the mice pups' brains may have been affected by the shaking of the mother's womb when she ran. Another possibility is that the chemical substances released when a mother mouse is running on a treadmill pass through the placenta and interact with the genes. Which would give the phrase "Born to run" a very literal twist.

THE SUCCESS OF THE KENYANS

Being a fan of physical exercise does not turn you into a good runner, of course, although it probably does help. If you go looking for the secret to speed, your search is likely to lead you to East Africa. Kenyans have dominated long-distance running since the 1950s, and Ethiopian runners have harvested their fair share of medals in that time too. Scientists are continually studying athletes from these countries in a bid to find the key to the success of the East Africans. What makes them so good? Many factors are cited as contributing to their success, including living at an altitude of six and a half thousand feet, where the air contains less oxygen. It is thought that because their ancestors have lived there for thousands of years, their bodies have adjusted to meet the demands of an oxygen-poor environment. When they compete at sea level, where the air is oxygen-rich, they are able to run like rockets. Other explanations include a low-fat and high-carb diet, a culture in which children are used to running a lot, and the motivation to escape difficult economic circumstances. In addition, there is the matter of genetic makeup. Most of the Kenyan runners who win race after race come from one of eight small tribes, collectively known as the Kalenjin. This makes it very tempting for scientists to trace the secret of their success to genes that appears to offer many anatomic and physiological advantages.

In 2012, Yannis Pitsiladis, a sports scientist and anti-doping expert at the University of Brighton, wrote a review of the evidence supporting the various theories. Curiously enough, to date no genes have been identified that can explain the success of Kenyan athletes. What we do know is that their success has nothing to do with the presence or otherwise of certain types of ACE and ACTN3, the two most-researched performance genes. Studies of maximal oxygen uptake, muscle fiber composition, diet, blood volume, and amounts of hemoglobin in the blood have also failed to identify the crucial factor behind their success.

We know that the speedy Kenyans and Ethiopians have a different physique from other professional athletes. They have longer and thinner calf muscles and Achilles tendons. The calf muscle is also higher and closer to the knee. This results in better running economy, as it costs them less energy to swing their legs back and forth.

Not only are their legs built differently, but the physiology of their muscles is also advantageous. In 1995, the Danish scientist Bengt Saltin at the Copenhagen Muscle Research Centre discovered that the muscles of East African runners burn fat more efficiently than their Scandinavian counterparts, which is a great help during a marathon (or a half marathon). It means they can maintain a fast pace without suffering too much acidification. The extent to which this can be attributed to training is uncertain, though. Some scientists believe that the amount of exercise they get during childhood may be partly responsible for the running talent of East Africans. Kalenjin schoolchildren start running at an early age, and not only as a sporting activity but also as a means of getting to school. The children get a healthy dose of aerobic training at a young age, and this appears to bear fruit later on. The same Danish research team also discovered that Kenyan teenagers who walk and run long distances to school have a maximal oxygen uptake (VO_2max) that is 30 percent higher than other teenagers. The VO_2max of these other Kenyan teenagers was equal to that of young Danish adults. Surprisingly enough, however, professional East African long-distance athletes do not have a higher VO_2max than European athletes. Ultimately, it is impossible to identify one single factor that accounts for the unparalleled success of Kenyan long-distance runners. It appears to be a mix of many different things.

MAXIMAL OXYGEN UPTAKE

Running magazines and blogs make frequent mention of the VO_2max, a term that is familiar to most runners. When you think fitness, you think VO_2max. It was an English physiologist, Archibald Vivian Hill, who discovered the link between oxygen and energy nearly a hundred years ago. The faster you run, the more oxygen you use up. Ever since this discovery, sports physicians have been using exercise tests to measure the maximal oxygen uptake of athletes in order to assess their stamina. The test is usually carried out on a bicycle ergometer or a treadmill, using a mask to measure the oxygen uptake and carbon dioxide production. When the oxygen uptake ceases to rise, you have reached your maximal oxygen uptake. Your VO_2max is a measurement of your (aerobic) stamina. It reveals how good your lungs are at extracting oxygen from the air you breathe in and mixing it with your blood, as well as the capacity of your heart and blood vessels to pump the oxygen-rich blood around your body and the capacity of your muscles to absorb the oxygen. The higher your maximal oxygen uptake, the better you are at using oxygen to release energy and the less likely you are to suffer from acidification.

When the ex-judoka and sports physician Jessica Gal was putting me through my paces, part of the routine was an exercise test on an ergometer. The test can help runners to create a more effective training program. Using a mask connected to a computer, Gal monitored the percentages of oxygen and carbon dioxide in my breath, with a series of electrodes also keeping track of my heart rate. In the meantime, the resistance level of the wheel was raised bit by bit. I was instructed to stop only when I could not continue anymore. My VO_2max turned out to be 42 ml of oxygen per kilogram of bodyweight per minute, a result that earned me a "good" score compared to the general female population and one that is well above the average. However, 42 ml/kg/min is still much lower than the average for a professional athlete. Well-trained endurance athletes can have a VO_2max as high as 90 ml/kg/min.

The charts that came rolling out of the computer also revealed that I was probably used to training just under or above my lactate threshold, the point at which acidification begins to set in, that is, when carbon dioxide production exceeds oxygen uptake. The VO_2max and the lactate threshold are interlinked: the higher your maximal oxygen uptake, the faster you must run to reach your lactate threshold.

PERFORMANCE PREDICTOR?

At age twenty-five and with a score of 42, I found myself in the "reasonably well-trained" category on the table for female endurance athletes. Sounds okay, but this was actually only one step above the "untrained" category. These terms are somewhat misleading, however, because your VO_2max score is related not only to how hard you train. Your maximal oxygen uptake is approximately 50 percent dependent on your genetic makeup, not to mention other factors like gender, age, and weight. I contacted the medical physiologist Erik Hulzebos at the UMCU to find out what your VO_2max score tells you exactly and what you can do to improve it. Hulzebos has written five books about exercise physiology, so I figured he was the best person to talk to. "You can improve your VO_2max by a maximum of 20 to 25 percent through training," he told me over the phone. The maximum amount of oxygen that you can take in while exercising says more about your talent for endurance sports than it does about the amount of training you do. "When the cyclist Lance Armstrong was having chemotherapy and was at his very lowest, he still had a VO_2max of 65 ml/kg/min, a score that many extremely well-trained athletes can only ever dream of achieving."

I took the exercise test in 2012 when I had already been running for six years. In 2013, I joined an athletic club and started taking my running a lot more seriously. Before I knew it, I was doing two interval training sessions a week, and since then my speed has increased significantly. It is reasonable to assume that this should apply to my VO_2max too. My Garmin smartwatch recently gave me a score of 51 ml/kg/min, although I don't know how reliable that is. Training has obviously helped to raise my score, but on the other hand I'm also a few years older now. The VO_2max of healthy but inactive people falls by about 9 percent every ten years after they turn twenty-five. However, you can reduce that to 5 percent every ten years if you take intensive exercise for the rest of your life.

According to my DNA test, I haven't been exactly lucky, genetically speaking, with my VO_2max. My DNAfit report tells me that my VO_2max potential is "average"; my body doesn't react very quickly to training, but not too slowly either. However, the DNAfit results were based on the data for only five genes, while scientists have identified at least 97 different genes that can predict how well you can train your VO_2max. But for the moment let's assume that my score is correct. What does it mean in terms of my athletic ability? To what extent can the score predict my race times? Thankfully, Hulzebos is able to answer my questions. A good VO_2max score is no guarantee of excellent race times. "You can't predict who is going to win a marathon based on VO_2max scores alone," he says. '"Your ability to make efficient use of the oxygen available to you, i.e., your running economy, is just as important as your ability to take that oxygen on board."

The Kenyans are living proof of this. Their VO_2max is no higher than that of European runners, but they are able to use it to generate more speed. "It all depends on how efficiently you release energy using the oxygen available to you. Your physique plays a role in this too. It's an advantage when you have less muscle mass in the calves and more in the hips and pelvic area." Running technique also has a significant effect on performance. For example, some runners use more muscles than they actually need to and waste precious oxygen on movements that make no contribution to speed. "A person with a high VO_2max and a poor running economy or technique often fails to perform as well as someone who has a lower VO_2max but who runs more efficiently," according to Hulzebos. Exactly the kind of encouraging news I was looking for. I might not be able to alter my oxygen uptake all that much, but I can definitely work on my running economy and technique. According to scientists at the Australian Institute of Sport, running economy—how efficiently you use your energy—can be a better predictor of performance for professional athletes than their VO_2max. However, there is no getting around the fact that without a high VO_2max you will never become a top marathon runner.

LATE DEVELOPERS

I have yet to run a marathon myself, primarily because of injuries. However, I plan to do so in the near future. Waiting until you are older to start running long distances can actually have its advantages. For recreational runners, your ability to run a marathon starts to diminish only around your fiftieth birthday, while for professional athletes thirty-five is the age at which performance starts to go downhill. The older we get, the poorer our physical performance becomes, partly because of the drop in maximal oxygen consumption and motivation. We also become less supple as our muscles and tendons begin to lose their elasticity.

These statistics are from a study carried out at Georgia State University and published in the journal *PLOS ONE*. In the study, physiologist Gerald Zavorsky and his colleagues examined the finishing times of runners in the Boston, New York, and Chicago marathons between the years 2001 and 2016. The fastest runners were all in the age category 25 to 34. The average age of the winners of the women's race was 31, while for the men it was 28. Zavorsky also discovered another interesting difference between male and female competitors. The times of the best male athletes slowed down by two minutes each year after the age of 35, while the best women's times deteriorated more dramatically: 2 minutes and 30 seconds slower with each passing year. For recreational athletes, however, there was no difference between men and women, with both sexes running the marathon 2 minutes and 45 seconds slower on average each year after they turned 50. An interesting question here is: why do recreational athletes' times begin to decline later than those of professional athletes? Zavorsky offers the following explanation: professional athletes make full use of their potential at an early age, while late developers often only start running after physical decline has already set in. And yet there is still lots of potential for improvement at an older age. If you have just taken up running in middle age, you can still improve and set personal bests for the marathon even if you were at your best, physically speaking, between the ages of 25 and 34. I have been inspired to start running marathons after the age of 30 by Fauja Singh, the British-Indian runner who ran his first marathon at the age of 89. In 2012, he ran the London marathon at age 101, finishing in a time of 7:49:21. The first person ever over the age of 100 to complete a marathon.

I am still improving my times for the 5K, 10K, and half marathon after ten years as a runner. It is, in fact, amazing that the records for the marathon are still being smashed on a regular basis and that the fastest sprinters on earth are capable of shaving tenths of seconds off existing records. But how far can we stretch the limits of human athletic ability?

FASTER AND FASTER

In 2009, Usain Bolt set a new world record by running the 100 meters in 9.58 seconds. The ten fastest times have all been run in the last twenty years and it seems inevitable that the world record will be broken again. Will the children being born today eventually dethrone Bolt? And will their children be even faster still? The former skater and cyclist Marije Elferink-Gemser is a human movement scientist at Groningen University, where she conducts research into talent recognition and development. What does she think? Are we approaching our human limits? "We are getting closer to our physical limits, but we're not there yet," she tells me on the telephone from the Netherlands' elite sports training center in Papendaal. What you do with your talent is the most decisive factor. A large proportion of our athletic ability has already been set in stone in the genes that shape our physique and personality. "However, we still see differences in performance development between people with similar DNA." The strange thing is that today's professional athletes are not much different from the athletes of yesteryear in terms of stamina. Their engines have not become any stronger. The fact that times continue to improve can be explained by better training practices, including working on running technique and getting more rest before a race. "We need to be smarter about our training methods. Not to do more, but to achieve more through training," says Elferink-Gemser.

Recreational runners are as guilty as anyone of training too hard and not getting enough rest before a race. Going all out used to be the philosophy in professional circles, too, but most coaches have since done a U-turn. Elferink-Gemser provides an example: "I used to skate with the Olympic champion Marianne Timmer. During training she only ever did six sprints, even though we were always told to do ten. Timmer just had a feeling that six was better in terms of optimal development. Later on, we found out that a skater's energy supply is exhausted after six sprints. Which begs the question: was there any point in doing four more sprints?" There is a lot of discussion about the benefit or otherwise of training long and hard. One thing it does for sure is increase the risk of overtraining and injury. "In our study we are following the development of over one thousand talented athletes. We have seen that those who have gone on to become top athletes already had a good idea at a young age what they needed to do to improve. They are highly motivated, believe in their own abilities, and do precisely what they need to do. And, unlike many lesser talented athletes, they take responsibility for their own failings and achievements; they are fully in control of their own development."

How much more headroom there is in terms of physical limits is different for each sport, according to Elferink-Gemser. Records for a relatively new sport are easier to break than those for running, which has been a high-performance arena for thousands of years. And then there's the role played by improved nutrition and advanced technology. There have not been as many technical developments in running as there have been in sports like skating and cycling, with the result that records appear to be stagnating. "Would you like to speak with Honoré Hoedt?" she suddenly asks me. "I just spotted him walking past." A minute later I am talking to the renowned Dutch running coach on the telephone. I can't believe my luck. Hoedt has coached many talented athletes during his career, including the middle- and long-distance world champion Sifan Hassan. So does he think that current records can still be broken? Records will always be broken, says Hoedt. "And that applies to all disciplines, although it does appear to be getting harder and harder to run the marathon in a faster time. Sprinting is still being overshadowed by doping too." According to Hoedt, a new record is simply a matter of waiting for the next genetically unique human to come along, someone like Dafne Schippers. "They will still be popping up twenty or thirty years from now." He also believes that place of birth is a very important factor. Without a program for identifying talent, some incredible athletes simply remain undiscovered. "If Dafne Schippers had been born somewhere in East Africa, she probably would not have made it to the top."

THE TWO-HOUR BARRIER

Ever since Roger Bannister became the first athlete to run the mile under four minutes in 1954—a feat that many had thought impossible—running the marathon in under two hours has been the holy grail of athletics. The official fastest time for a marathon—2:01:39—was run in 2018 in Berlin by the Kenyan Eliud Kipchoge. To shave one minute and forty seconds off that incredible time he would have to run each kilometer almost two and a half seconds faster.

In the spring of 2017, Kipchoge and two fellow athletes, Zersenay Tadese from Eritrea and Lelisa Desisa from Ethiopia, decided to try to break through the magical marathon barrier at an event in Italy. The attempt, labeled Breaking2, was organized by Nike, which had put together a team of scientists to work on the project. The athletes were not running in an official marathon but in a private race on the Formula 1 circuit in Monza. The men had trained hard, but experts believed that the greatest gains would be because of the ideal conditions.

Breaking the two-hour barrier was possible, theoretically in any case, said scientist Wouter Hoogkamer of the University of Colorado in an article in *Sports Medicine*. Together with two colleagues, one of whom was a paid advisor for Nike, he calculated what was required to run a time of 1.59.59. The most room for improvement lay in aerodynamics. There were two ways in which a total of three minutes could be shaved off. The first involved the use of a strategy called "drafting", where the runners in front create a slipstream that pulls the runners behind along in their wake. The designated athletes would run the first half of the race behind a group of pacesetters. In the second half they would take turns running at the front, like cyclists do. It is impossible to profit from both a following wind and drafting, as a following wind cancels out the effect of running behind another person. "A lot of time can be gained from drafting, however, because it means that the athletes can stay sheltered from the wind for the entire 26 miles," explains Hoogkamer. "But it is also one of the reasons why the International Association of Athletics Federations (IAAF), now known as World Athletics, will not recognize a record run under these conditions." Basically, using rotating pacesetters is not permitted.

The second way to save time revolves around footwear. Every extra four ounces of shoe costs the wearer 1 percent more energy. Hoogkamer based his calculations on the world record time of 2:02:57 run by the Kenyan Dennis Kimetto in 2014. Kimetto set the record in shoes that weighed eight ounces each. Hoogkamer's calculations suggested that wearing lighter shoes could shave another minute off the record. For the record attempt in Monza, the three athletes wore a shoe specially designed by Nike: the Vaporfly. According to the manufacturer, the secret to this shoe lies not in its weight but rather in the material used in the midsole, which is lighter and springier than the widely used foam. The sole also contains a curved carbon fiber plate that helps push the runner along. Nike claims that the Vaporfly makes you run 1 to 4 percent faster. Kipchoge came agonizingly close to breaking the two-hour barrier in Monza, falling short by only 26 seconds. His time of 2.00.25 still represented a phenomenal performance and, at the time, an unofficial world record.

Almost everyone knows what happened next. In the spring of 2019, Kipchoge got his revenge in Vienna. This time the conditions were even more favorable. Teams of pacesetters (forty-one in all) ran in a V formation in front of Kipchoge to protect him from the wind. The car driving in front shone a green laser light on the road indicating the ideal route through the Wiener Prater park. Thousands of supporters from all over the world lined the route cheering him on. And he was wearing an updated version of the Vaporfly: the Alphafly. This time, all the ingredients combined to help Kipchoge achieve the impossible and he broke the two-hour barrier in a time of 1.59.40. Eliud Kipchoge was already considered the best marathon runner of all time, but his performance in Vienna made him immortal.

Michael Joyner, an expert in human performance at the Mayo Clinic, had thought that it would take longer to break the barrier. In an article in *Journal of Applied Physiology* he estimated that, based on the physiology and genetic makeup of athletes, the magic barrier would not be broken before 2021 or 2022. His figures showed that since 1960 the record had fallen by an average of twenty seconds each year. If the trend had remained the same, it should have taken longer to dip under the two-hour barrier. He also predicted that an East African athlete—one who is used to high altitude and very physically active—would be the first to achieve the breakthrough, a prediction that proved correct.

Joyner first started speculating in 1991 about the possibility of running a marathon in under two hours. It wasn't long before other scientists joined the discussion. In an email exchange, he tells me that the largest gap in our knowledge has always been in the area related to the efficiency of running. And this is still the case, he says. "The shoes developed by Nike were designed with more efficient movement in mind. They endeavored to make tiny gains in a number of areas: no more straight edges; the use of pacesetters; drafting; footwear; nutrition; and good weather." According to Joyner, breaking the two-hour barrier does not necessarily require a scientific approach. It is primarily a matter of the right runners running on the right day and on the right course. And with the right kind of financial incentive, which, incidentally, was not Kipchoge's main motive. The next step is for an athlete to run an official sub-two-hour marathon, and the rumor around town is that Kipchoge is aiming for that honor too.

AIMLESS AMBLING

Up until now, we have been focusing primarily on the physical side of athletic performance. The idea is that your level of lactic acid and your VO_2max are what define the limits of your athletic ability. If your muscles begin to burn and your heart cannot supply them with enough oxygen, then it's time to quit. Or not? When you think carefully about it for a moment, you quickly realize that this doesn't quite add up. There has to be more to it than that. When you watch runners during the last stages of a race, you see that they are always able to squeeze out a final sprint, even after a long, hard run. I've experienced this myself too. No matter how exhausted I am, I am usually able to accelerate at the end of a race. Amazingly, there's often more than enough power left in my thighs. How is this possible when my muscles are screaming with pain? Surely the increased acidification should only slow me down.

This final sprint is possible because it is the brain and not the heart or muscles that calls the shots when it comes to physical exercise. It is the brain that detects the build-up of lactic acid and other metabolites and translates this into a burning sensation. And when the supply of glycogen begins to diminish, it is the brain that makes you feel like you've hit the wall and that creates one illusion after another in an all-out effort to get you to throw in the towel. So does that mean it might be possible to postpone your demise if you are blessed with a healthy portion of mental resilience? Motivation and discipline, as every coach and athlete knows, have a crucial effect on athletic performance. A high VO_2max will certainly help you to run faster, but your psychological skills are what determine how much you can draw from your body when the going gets tough. Are there ways in which you can improve your resilience? I certainly hope so. In fact, I am quite sure that the average runner can make a lot of gains on the mental side of things. Time to dig deeper into the psychology of performance.

TIP

A better running economy will boost your performance. You can only alter your physique by reducing your body's fat percentage, and your VO_2max can only be trained to a certain degree. However, you can adjust your running technique. It might be a good idea for you to join a running group or club.

10

FATIGUE IS ALL IN THE MIND

Motivation is a skill. It can be learned and practiced.
—Amby Burfoot, winner of the 1968 Boston Marathon

A few years ago, scientist Ashley Samson embarked on a project aimed at accessing the darkest recesses of the runner's mind. What goes on in the minds of people who voluntarily expose themselves on a regular basis to the rigors and stress of long-distance running? Samson is attached to California State University and also runs a private clinic for athletes who wish to avail themselves of her expertise as a sports psychologist. Samson was an athlete herself in her younger years and she still runs ultramarathons, so she knows all about the mental trials of running.

Up until recently the only way to get inside the heads of long-distance runners was to ask them to fill out a questionnaire after a race. Not exactly what you would call a reliable method, as it is always uncertain how well people remember specific information after the event. Samson and her colleagues decided to try something different. They fitted ten runners with microphones and asked them to articulate their thoughts freely and without any self-observation while out on a long run. The scientists then listened to all eighteen hours of the recorded material, searching for patterns. The thinking-aloud protocol allowed only immediate thoughts to be recorded; thinking aloud actually stops the mind from wandering. Nevertheless, the scientists must have had great fun listening to the recordings. "Holy shit, I'm so wet [from all the sweat]," reported Bill. "Breathe, try to relax. Relax your neck and shoulders," said Jenny. Bill found the going very tough: "Hill, you're a bitch . . . it's long and hot. God damn it . . . mother eff-er." Fred paid more attention to his surroundings: "Is that a rabbit at the end of the road? Oh yeah, how cute."

Samson categorized the thoughts into a series of themes. Three themes in particular emerged: pace and distance; pain and discomfort; and environment. All of the participants in Samson's experiment experienced some level of discomfort, especially at the beginning of their run. For example, they suffered from stiff legs and minor hip pain that became less severe the longer they ran. To cope with the pain and discomfort, the runners used a variety of mental strategies, including breathing techniques and urging themselves on.

There is more to running than just training your muscles and improving your stamina. It is also a mental sport, and maybe even more so than previously believed. I found this out the hard way on Easter Monday 2017 when I entered a half marathon being run on the castle grounds near the Dutch village of Haarzuilens. I spent the first twelve kilometers enjoying my surroundings: the gardens, the lake, and the refreshing wind in my face. At around fifteen kilometers my legs began to feel a bit heavier, but I was still able to maintain my pace of five minutes and ten seconds per kilometer (eight minutes and twenty seconds per mile). A quarter of an hour later my whole body was hurting. My legs were tired, but I kept going. I struggled on, but the nineteenth kilometer was so tough it took me five and a half minutes to complete it. All I could think of was the burning sensation in my legs and hips, of how awful it felt. At the twenty-kilometer mark I managed to suppress these negative thoughts and I picked up my pace again, though I still don't know what hidden energy reserves I was able to tap into. I turned a corner and there were my parents and boyfriend cheering me on. I used the extra energy to accelerate again. My last kilometer was the fastest by far and I finished well inside my target time of one hour and fifty minutes.

Most runners appreciate the importance of mental strength. Men and women who decide to join their colleagues for a 10K run without any prior training are often able to show just how far you can get on motivation and perseverance alone. They run on "mental energy" and spur each other on. Keep going! Never mind the pain! As for ultramarathon runners, instead of ignoring pain they embrace it as part of the whole experience of long-distance running. "Pain is inevitable" is their mantra; it is an essential ingredient of the running experience. So what are the psychological qualities that make you a good runner? To what extent do they influence performance? And most importantly: can you train mental toughness?

THE PSYCHOLOGY OF PERFORMANCE

Anyone who wants to know more about the psychological side of sports would be well advised to talk to Vana Hutter.

She is an expert on the mental health of top-class athletes, and she sums up all of the research on the matter as follows: top-class athletes are armed with high levels of self-confidence, dedication, and focus, as well as the ability to concentrate and handle pressure. Their academic performance and social skills are also often better than that of nonathletic types. According to Hutter, athletes need self-regulation in order to perform. Everyone can learn, to some extent at least, to control their emotions, thoughts, and actions. And it is this aspect—learning to self-regulate—that is of particular interest to runners.

Funnily enough, Hutter began her scientific career at the "hardcore" end of exercise physiology: physical measurements of athletes' bodies. "As time went on, however, I realized that athletic performance is determined by a combination of body and mind," she tells me over coffee in Amsterdam. "I discovered that it is far more difficult to predict athletic performance than some physiologists would have you believe. There are so many factors that we just can't account for." For example, how do you explain the fact that the times athletes run are so different despite their being physically very similar? If you were to subject the top ten marathon runners to a physiological examination, they would probably all have a high VO_2max and excellent running economy. Some top athletes have something extra as well, however. "Measured over a longer period, the trainability of athletes is more or less the same. What really matters during competition is the extent to which their physiological systems are primed and ready to go, and how well those systems cooperate with each other," explains Hutter. "Whether an athlete can avail of their maximum physical potential at the crucial moment is partly a mental matter."

She provides an example. "If your muscles are a little bit more tense because you are nervous, this will have an effect on your movement efficiency. You will need more energy to achieve the same kind of forward motion. This is the biomechanical explanation of the role of psychology in performance. On the other side of the spectrum, nervous anxiety can result in negative thoughts and fear of failure." In other words, to go far as an athlete you need not only the right kind of physique but also to be mentally strong, primarily because of the influence the psyche has on how the physical body performs. Mental strength may in fact be the thing that separates the winners from the rest of us. Today, no one denies the role played by psychology in athletic performance. However, the extent to which coaches address mental toughness when training their athletes is a different matter, according to Hutter. Most of them do integrate it in their training, but opinions vary greatly on just how trainable mental toughness actually is.

SELF-REGULATION

What makes you "mentally tough"? What does it require you to do? Or indeed not to do? Sports psychologists still haven't come up with a clear answer. Mental toughness is a catch-all term without any well-defined meaning, explains Hutter. "We associate mental toughness with the ability to deal with difficult situations. And it helps if you are armed with a wide range of coping mechanisms, as well as the creativity required to turn difficult situations to your advantage." In any event, one thing you really need in order to train and perform well is self-regulation. Perseverance, the ability to block out your surroundings, clear goals, and being able to cope with stress are the skills associated with self-regulation. "We know that the capacity to self-regulate is a quality that sets great athletes apart. It's a weapon you need to have in your arsenal, but also one you need to learn how to use."

There are two of kinds of self-regulation, and they are often used interchangeably in scientific literature. The first is self-regulated learning, which is important in every kind of sport. It involves taking control of your own development process and using every available opportunity and situation to keep on improving, for example by tackling the steep hill instead of sticking to the flat track or going to training after a hard day's work or a bad night's sleep. "Tackling difficult situations head-on teaches you more about your own capabilities and that it pays to persevere even when you are exhausted."

The second kind of self-regulation concerns how to control your emotions, thoughts, and actions and keep them in line with your goals. For example, how do you deal with the inevitable nerves before a race and feelings of boredom and fatigue while you are running? "Some people have a natural talent for self-regulation," says Hutter. "Even children can be very good at it from a young age." She cannot say for sure, however, whether top-class athletes are born with an inherent talent for self-regulation or develop it from practicing their sport. "Self-regulation can be learned to some extent, but we do not know how trainable it is, primarily because of its complexity. I think there's a limit to its trainability. People who are very bad at it can certainly improve. But they will probably never be as good as those who have a natural talent for self-regulation or have worked on it from an early age."

So how should recreational athletes train their self-regulation? Should they employ a coach or sports psychologist? "A sports psychologist can help, of course, but some simple background information is usually enough to get you started," Hutter tells me. "You need to actively seek out situations in which you are forced to confront your own thoughts and emotions. That has the most effect." We may not always realize it, but every time we train we are exposed to a lot of different psychological stimulants. "We all require motivation to complete a training session. Sometimes you have to dig very deep to find it, and sometimes it's there at your fingertips. Increasing your pace and pushing on through the fatigue is a form of mental power training. Even just making time for an endurance training session lasting a couple of hours involves a psychological process."

AMBITIOUS BUT REALISTIC

Goals inspire action, and most marathon runners are good at setting goals, according to a study carried out by the National Bureau of Economic Research. A few years ago, a team of behavioral scientists at the bureau examined the data collected from millions of runners. Most of them were capable of completing a marathon in four hours, as you would expect. One finding that raised a few eyebrows was that the finishing times around 3:58 far outnumbered those closer to 4:02. It appears that most runners finish just under the hour or half-hour mark, partly because of the pacesetters pulling them along, so that there are always far more finishing times recorded in the vicinity of 3:59 and 4:29. The hour and half-hour marks serve as good reference points for runners, the scientists argued. It gives them something to aim at and helps them to find the (hidden) energy required to achieve their desired time. Runners associate the minutes and seconds over the hour and half-hour marks with failure and everything under with success.

So having a clear goal can be a great motivator. Imagine you are about to run a marathon for the first time. Should your goal simply be to "make it to the finish line" or is it better to set yourself a target? The former is generally regarded as the best option for your first marathon. However, Hutter believes that this is different for every individual. "My advice is to set yourself a goal that will motivate you to keep going, but at the same time is comfortable enough for a novice marathon runner. For some people, merely making it to the finish is not enough, but it's a good place to start. As a goal it can help to make you less nervous about the task at hand, but it does not provide much in the way of motivation. The perfect goal is one that you can fall back on while simultaneously pushing you forward."

The problem with setting a target is that you may spend all your time looking at your watch and suffer under the pressure of trying to achieve your goal. Enjoying the run is then completely out of the question. '"You can also switch between strategies: run according to a set schedule while also slowing down every now and then to recharge the battery and enjoy what you are doing. The most important thing is to learn how to be happy with a gentle pace and to trust it as the best pace for you. And not to give in to the impulse to up the tempo. Suppressing the urge to run faster is all about regulating your emotions, and that is something you have to train."

Long endurance runs are perfect for this kind of training. They allow you to learn more about what goes on inside your head and body when you are running. "Learning how to deal with nerves, cope with rising fatigue, pace yourself when you still have a long way to go—that's the psychological function of training." It is also useful to set short-term goals that are easily attainable in addition to your ultimate long-term goal for a marathon. "Goals that you can achieve during one training session, like 'today I'm going to run fifteen miles,' can be an excellent source of motivation. Achieving these short-term goals is good for your self-efficacy: the conviction that you are capable of achieving a certain goal. Each specific goal that you manage to achieve feeds that conviction. And the greater your self-efficacy, the more ambitious your goals can become.

Imagine setting yourself the goal of running your first marathon in under four hours. What follows if you don't succeed? Disappointment. "Setting yourself an ambitious goal probably means you have plenty of self-confidence and a desire to be motivated," says Hutter. "And if you fail to achieve that goal, you probably have enough mental toughness to deal with that too. In fact, failure can even motivate you to make sure you succeed next time around, as long as your self-confidence hasn't been shattered in the meantime."

THE MYSTERY BEHIND STOPPING

There are limits to human athletic ability, regardless of how well trained you are or how many mental strategies you have

at your disposal. Although it differs for each individual runner, eventually we all reach a point where we have to give
up. Within the realm of sports science, physiologists and
psychologists are all looking for the answer to the question:
what causes us to stop or slow down during a race? After all,
at the moment when we stop we usually still have enough
energy in the tank. The decision to stop running has nothing to do with your muscles or energy system and everything
to do with your brain. Experts are in unanimous agreement
that it is the brain that controls physical exercise. However,
they are still arguing about how it persuades us to stop before
we reach the point of complete exhaustion. Does the brain
act on signals from the body, or is it our psyche that pulls the
strings? The question has given rise to a fascinating theoretical discussion.

One of the liveliest contributors to that discussion is
Samuele Marcora at Kent University in England. He believes
that the reasons for fatigue while running are of a purely
psychological nature. His research suggests that signals from
the muscles, heart, and lungs do not play a significant role
in the decision to stop or slow down. Psychological factors,
however, such as mental tiredness after a day spent staring
at a computer, do have a direct effect on the decision to stop.
Marcora is one of the best-known scientists studying the perception of exercise among endurance athletes. In his opinion, what runners refer to as exhaustion has nothing to do
with their physical ability to carry on or not. It is simply a
matter of deciding to give up.

Marcora's official title is Professor of Exercise Physiology,
but he feels more of an affinity with psychology than with
physiology. Sport and exercise are goal-oriented behaviors

that are fueled by motivation. And, as he explains, the branch of science that studies behavior is not physiology but psychology. I attended a lecture given by Marcora at Radboud University, where he explained his concept of fatigue. Afterward we sat down for a chat at a picnic table on the university campus.

The focus of his research is on fatigue in endurance sports. Marcora is trying to establish why humans are unable to maintain a certain speed or level of strength indefinitely. What causes us to slow down, sometimes even to walking pace, during a race? "Up until very recently it was assumed that a person could continue exercising until they reached the point where their body was unable to transport enough oxygen to the muscles," Marcora tells me. "In that case, the muscles are no longer able to generate the required power quickly enough. They are just too tired." The generally accepted theory was that the body always reaches a point at which it has to stop, regardless of how motivated you are. However, there has never been any convincing data to support that model. Marcora believes that we rarely reach the point of physical exhaustion while running. The results of his own research contradict the idea that we stop running as soon as we receive certain signals from our body.

In 2010, Marcora and his colleague Walter Staiano invited ten male athletes to their lab for an endurance test in which they were asked to pedal for as long as possible on a bicycle ergometer set to a certain level of resistance. Before the test started, Marcora and Staiano asked each athlete to pedal as hard as they could for just five seconds. A record was kept of the power generated by their leg muscles. After this short, explosive test the men were asked to cycle for as long as possible until they couldn't carry on. The average time was twelve minutes. It was the final part of the test that proved the most interesting. After the endurance test the scientists asked the athletes to repeat the five-second explosive burst of cycling. Just picture it: you are completely exhausted but you are asked to cycle like a madman again. Surely your legs would refuse. Nothing of the kind, as it turned out. The men did not score as well in the second explosive test as they had the first time around, but they were still able to generate three times more power than they had during the longer endurance test. Isn't that strange? First you give up because you can't pedal on anymore, only to deliver another explosion of power immediately afterward. This means that tired muscles and a lack of energy are not the problem, according to Marcora and Staiano. So what caused the cyclists to give up? Motivation, or rather the lack thereof, they suggest. The participants knew that the last test would only take five seconds and so were able to come up with the goods. The endurance test, on the other hand, lasted much longer, without the athletes knowing precisely how long they would have to keep pedaling. This is probably what caused them to lose their motivation. "When someone stops because they are exhausted, they still have plenty of energy left over," says Marcora.

It's a different story for more explosive sports, however. In the case of weight-training, there is a point past which your body cannot go on. After a certain number of push-ups, your muscles simply cannot generate enough power to continue. Instead, your arms begin to tremble and you collapse to the floor. Kevin Thomas and his colleagues at Northumbria University in England conducted an experiment with cyclists in which they demonstrated that the shorter the period of physical exertion, the more exhausted the muscles become. And the longer the period, the more tired the brain becomes. So in the case of short, intensive exercises, the legs suffer the most, while longer endurance exercises tend to exhaust the brain.

MIND OVER MUSCLE

In 2012, the renowned South African sports scientist Tim Noakes also questioned the idea that burning muscles are the dominant factor when it comes to our ability to carry on. He believes that our brain houses a kind of subconscious command center (which he calls the "central governor") that protects our body from damage like extreme exhaustion or torn muscles. This command center monitors the incoming signals from our body, such as the level of metabolites in the muscles and the body's supply of sugar. If the risk of damage is acceptable, we can carry on running. However, the command center always pulls the plug and tells us to stop long before we've used up our energy supply. Noakes believes that its job is to ensure that we never go beyond our physical limits and do real harm to ourselves in the process. The central governor theory is well known among scientists, but Marcora is not a big fan. He believes that it assigns too important a role to the signals received from the muscles, heart, and lungs. But how does our brain make us stop if it doesn't make use of the signals coming from our muscles? "I am not saying that what happens in the body is of no consequence," he says. "But the all-important factor in the case of endurance sports is *perception of effort*."

"Perception of effort" is a subjective feeling that one might express as "Oh boy, this is tough going." Runners constantly try to find the right balance between the maximum amount of effort they are prepared to put in to achieve their goal and the effect that effort has on them. Imagine you have set yourself the goal of running a half marathon in under two hours. For the first ninety minutes you have no problem maintaining your pace of 6.5 mph, even though the run is feeling tougher as you go along. That feeling continues to grow stronger until you reach a point where you are so exhausted that you cannot carry on. The feeling of exhaustion is greater than the amount of effort you are prepared to put in. The result? You slow down. In fact, you might even throw in the towel and walk the rest of the way.

"If you run at the same pace for a long period of time, the perception of effort increases as you go along," continues Marcora. "It feels increasingly harder to keep on running, even though your muscles are providing the same amount of power at the same speed on a continuous basis. At a certain moment, however, the perception of effort reaches a maximum value that forces the athlete to stop. Even the most motivated athletes have to give up at this point, the point of exhaustion. And that's the kind of 'fatigue' I'm interested in."

He still doesn't know exactly where this perception of effort comes from. It might even be the organs that relay the information. Studies have shown that perception of effort is closely related to how fast your heart beats. However, this doesn't mean that a high heart rate is the source of the feeling. It might even be the other way around: that the heart beats faster because the physical exercise feels very demanding.

Suddenly Marcora stands up from the picnic table. "Let's do an experiment," he says, and then asks me to stand opposite him. "Lift me up and tell me how difficult it is." This is not going to be easy, I quickly realize, because he is not exactly small. I put my arms around his waist, count to three, and then summon every ounce of strength I have in me. I can't lift Marcora one millimeter off the ground. I let go. "Not easy to lift 275 pounds, is it?" he says grinning. "On a scale of one to ten, how would you rate the effort it required?" he asks. "Ten," I say. I had given it everything I could. "And did you feel exhausted?" No, I did not experience any feelings of fatigue. The point he is trying to make is that a high perception of effort is not the same as feeling exhausted. "Fatigue is a rather vague concept, and one that I prefer to avoid," says Marcora. "Fatigue is actually a frame of mind, a lack of energy. You can feel fatigued even when you are sitting down doing nothing. Perception of effort, however, is not a frame of mind. It is a feeling you only experience when you voluntarily contract your muscles." His position is crystal clear: exercise is all about perception.

MENTAL FATIGUE

Marcora and his colleagues carried out an experiment in 2009 in an attempt to prove that the perception of effort is what causes us to stop exercising. Sixteen participants were invited to their lab, where they first filled out a questionnaire related to their mood at that moment. They were then asked to sit in a dark room, where one group of participants was given a difficult computer assignment that lasted ninety minutes. A computer assignment requires cognitive activity and therefore taxes the brain; it makes you mentally tired. The other group—the control group—was told to watch a documentary about cars and trains; they experienced no mental fatigue. When they emerged from the darkened room the participants were once again asked to fill out a questionnaire describing their mood, and to answer an extra question related to their motivation for the next part of the experiment: a cycling test.

The men and women taking the test were instructed to sit on a bicycle ergometer and were fitted with a mask to measure their respiratory gas exchange and electrodes to monitor the heart. They were then told to pedal as fast as they could until they could pedal no more, with the resistance being increased every two minutes. To provide extra motivation, there was a prize of $50 on offer for the cyclist who could last the longest. During the test, research assistants asked the cyclists at regular intervals to rate their perception of effort on a scale of one to fifteen. "The only way to measure perception of effort is to ask people to gauge the effort demanded of them," according to Marcora. After the cycling test, the participants filled out the mood questionnaire for the third time. Everyone was asked to return to the lab for a second session in which the participants who had watched the documentary were given the computer assignment instead, and vice versa.

The difference was crystal clear. The test subjects who had to apply their cognitive powers during the computer assignment caved in more quickly during the subsequent cycling test. They also rated the difficulty of pedaling on a lot higher than the control group. The funny thing is, it had nothing to do with their heart, lungs, or muscles, which continued to function perfectly according to the data from the mask and electrodes. Furthermore, the computer assignment had no effect on the level of lactate in the blood, and the VO_2max was more or less the same for both groups.

Where the groups did differ was in the levels of mental fatigue. The results of the questionnaire revealed that the brains of those tasked with the computer assignment were a lot more tired before they took the cycling test. However, they were not less motivated. While the cycling test became progressively more difficult for both groups, the participants who were mentally fatigued reached the maximum level of effort they were prepared to put in much quicker before quitting. Conclusion: a cognitive computer assignment has no effect on your muscles, but is does exhaust you mentally, which in turn has a negative effect on your endurance performance. Mental fatigue increases the perception of effort, that is, your perception of how hard it is to keep going. In 2017, a group of Dutch and Belgian scientists published an overview in the journal *Sports Medicine* of the studies carried out into mental fatigue, all of which suggested that mental fatigue has a negative effect on endurance performance. So it appears that if you are mentally fatigued, you are likely to throw in the towel a lot sooner.

FAT IS GOOD

The results of Marcora's experiments are impressive and his theory of perception is convincing. He is currently working on a new theory on the evolution of the perception of effort. Why did humans develop this psychological mechanism? The mechanism is not unlike the command center in Tim

Noakes's model, which acts as a defense mechanism and prevents us from causing serious harm to ourselves by reacting to the signals it receives from the body. According to Marcora, the human perception of effort has also evolved, but for a completely different reason: to keep us fat.

"Running often and for long periods of time makes you thinner," he said during his lecture at Radboud University. "On the African savanna, where food is scarce, it is not a good idea to lose too much fat because it reduces your chances of survival. Early humans were only able to generate a limited amount of energy from their food, so it was important not to expend that energy unnecessarily." His theory is interesting, and it gels nicely with the theory proposed by biologist Daniel Lieberman, who believes that humans evolved to be able to run long distances. Lieberman also says that humans evolved to avoid unnecessary activity. "Making running feel increasingly more difficult, so that you eventually have to stop, prevented us from depleting our fat reserves."

How long would a human be able to run in the absence of mental obstacles and without experiencing the feeling that they can't carry on? "Who knows? Days, maybe. I think that the loss of fluids and low blood pressure would cause you to faint first before your muscles lost the ability to contract for want of energy. Again, energy expenditure is almost never the problem," says Marcora.

Some scientists have difficulty accepting perception of effort as a limiting factor in running. They claim that perceptions are merely observations to which we assign meaning, and that we can choose to ignore them if we wish. Physical signals, however, like tired muscles and a lack of energy, are more difficult to ignore. Marcora admits there is a limit to the extent to which you can overcome perception. "But the fact that something is perceived does not make it any less real or forceful. The perception of effort really makes you feel like you cannot go on."

If this exercise physiologist's theory is correct—that perception of effort is the factor that limits our ability to carry on—we should actually regard it as good news. There is strong evidence to suggest that psychological manipulation, such as lowering the perception of effort, can help boost endurance performance. For sports psychologist Vana Hutter the matter is less clear-cut, however. In her opinion, neither Noakes's nor Marcora's model reveals the whole truth. "We just don't understand the full complexity of it yet. While we know that certain mental strategies can aid performance, we still have no idea which buttons in the brain we are pressing." It is also debatable just how much difference it makes whether endurance performance is regulated by perception of effort or by a command center in the brain. What we do know for sure is that psychology is an integral part of sport. "Humans are not robots. Even if the science of physiology eventually comes up with the perfect training program, it is humans who will have to put it into practice. And behavior is never free from psychology."

MENTAL TRAINING

It's 7 a.m. on Monday morning and my alarm has just gone off. I don't feel like getting up. I had a late one last night and I'm still tired. But up I get, casting a jealous look at my boyfriend as he turns over for another hour's sleep. He doesn't have a ten-kilometer run on his schedule before the working day begins. I gobble down a banana before heading out the door. After only three kilometers I'm already beat; it feels really tough today! My legs are refusing to cooperate, I'm gasping for breath, and all I can think of is the string of kilometers I still have ahead of me. Then it starts to rain. And yet these are precisely the conditions I was hoping for, because I know that we can train our brain to get used to feelings of fatigue.

Brain training is not unlike regular training. When you start running for the first time, your legs soon grow tired and you are quickly out of breath. The more you train, however, the more your body adjusts: tendons, bones, and muscles all become stronger and your stamina increases. To make your brain stronger you need to do some tough mental training, like going for a run after a hard day at work. This helps you to delay the point at which running begins to feel really tough. It's all about developing the kind of mental resilience that comes from going out for a run even when you really don't feel like it. Luckily there is no shortage of potential tough-going scenarios, including setting your alarm for an early morning jog after a night out on the town.

If there is one thing that shatters you mentally, it has to be too little sleep. Even just one bad night's sleep is enough to undermine your performance, because the effort you have to put in feels much heavier. This probably explains why the top Belgian swimmer Pieter Timmers had his own mattress flown to the Olympic Games in Rio de Janeiro in 2016. His performance at the European Championships a few months earlier had been disappointing, and he attributed this to sleeping poorly.

Sleep is crucial to effective cognitive and physical performance. Science has yet to come up with an explanation for the exact function of sleep, but there are many theories on how it strengthens the immune system, replenishes the energy we consume during the day, and stimulates our cognitive development, such as learning and memory skills. Various studies have shown that when adults get less than seven hours sleep, they perform worse at tests on alertness, reaction time, memory, and decision making. A lack of sleep also has a negative effect on your mood, which can undermine an athlete's motivation and dedication to their sport.

That said, training when suffering from a lack of sleep is a good thing to do every now and then because it makes you mentally tougher, which in turn can help you run for longer periods of time. However, before a race you should always get enough sleep to avoid feeling mentally fatigued at the starting line.

MUSIC

Music is known to put some pep in your step, but it can have a calming effect too. In scientific literature you will find the story of a long-distance runner who selects the music he listens to depending on his emotional state. An hour before a race he listens to slow, relaxing songs so that he can remain calm. Just before the race starts, he injects a bit of pace into proceedings by selecting faster songs with inspiring lyrics. He also adjusts his music to match his speed, picking songs with a relatively low bpm (beats per minute) at the starting line to ensure that he doesn't set off too quickly. Basically he uses music to create the right mindset. Science agrees that motivational music can inspire pleasant emotions, like a sense of calm or high energy, when you are running. It can also help to temper unpleasant emotions like fear of failure and fatigue, according to a study involving sixty-five volunteers carried out at the University of Wolverhampton.

In 2014, two scientists at the Brunel University in London published an update on their research into music and sport in the *Sport Journal*. One of them, the sports psychologist Costas Karageorghis, is a widely respected authority on the effect of music on athletic performance. He provides advice to athletes and has published a book called *Applying Music in Exercise and Sport*. According to Karageorghis, listening to music can be hugely beneficial. One of the benefits is that music can trick the mind into believing that you are less exhausted than you actually are during a workout. Music distracts you from thinking fatigue-related thoughts, which tempers the subjective feeling associated with the effort you are making. However, this trick doesn't work every time. Music has a psychological effect only when your heart rate is not too high, that is, during light to moderate exercise. As soon as you start gasping for breath, all the music in the world won't help you suppress feelings of fatigue, because your brain will simply have stopped listening.

However, a study carried out in 2008 at Florida State University suggested that music can be beneficial during strenuous exercise too. Fifteen male students were asked to take an intensive test on a treadmill. The task was simple: run at a fast pace until you can run no more. Each student did the test four times: once while listening to rock music, once to dance music, once to "inspirational" music, and once without any music at all. None of the different types of music helped the runners last any longer, but one third of the students said that the music did help them at the start of the run and inspired them to keep going.

The British athlete Paula Radcliffe knows how helpful music can be during training. Karageorghis quoted her in the article in the *Sport Journal*: "I put together a playlist and listen to it during the run-in. It helps psych me up and reminds me of times in the build-up when I've worked really hard, or felt good. With the right music, I do a much harder workout." One of her favorite tracks is "Stronger" by Kanye West. Probably because of the song's beat, although the lyrics probably provide extra motivation. As for me, I like to listen to my guilty pleasures from the '90s, specifically the happy hardcore stuff. When I start to get tired I enlist the help of Charlie Lownoise and Mental Theo. Happy hardcore accompanied me all the way to the finish line the day I ran my first 10K.

Apparently, it is also a good idea to listen to synchronous music—tracks whose rhythm matches the pace at which you run. Costas Karageorghis demonstrated this in an experiment with a group of athletes who were asked to synchronize their pace with a specific number of beats per minute. It had a stimulating effect: they continued running for 20 percent longer than in a test without music. The music doesn't even need to be of the motivational kind; even if the songs are boring, they can pep you up as long as they are synchronous with your pace. Many running apps feature Spotify as standard, and the music you listen to is automatically adjusted to match your pace. The app calculates the number of steps you take per minute and then picks tracks with the same bpm. Now that's what I call handy.

The idea is that the rhythm of the music helps you to maintain a steady pace. And running efficiently requires less energy, which means you can run for longer. Synchronization works the same way in other sports that require constant repetition of the same motion, like rowing and cycling. In a study carried out in 2008, a group of test subjects took part in a cycling test. Those who listened to synchronous music while cycling required 7 percent less oxygen for the same amount of effort than those who listened to music whose rhythm was slightly slower than their pedaling.

To sum up: music has proven itself to be a useful tool for athletes who want to train harder, maintain a steady pace, control their emotions, reduce perception of effort, and even lower oxygen consumption. Nevertheless, at the New York City Marathon they prefer to see runners without music in their ears. In 2007, the organization tried to ban earphones and portable music devices like iPods for security reasons. They claimed that runners listening to music are unable to hear announcements on the course and are less aware of other runners around them. This is not a problem for professional athletes, as their priority is to focus on their body and their fellow competitors, not on their favorite music. It's a different story for most recreational runners, however, who often need their carefully compiled playlists to keep themselves motivated. In the year when the ban was first introduced, hundreds of runners simply ignored it and brought their music along. One runner told the *New York Times* that she would prefer to be disqualified than hand over her iPod. Today the organizers continue to advise strongly against listening to music during the marathon, but they usually turn a blind eye to the practice. Partly because it would be impossible to check every single runner.

BRACE YOURSELF

There are lots of other psychological tricks that can have a direct effect on the perception of effort. Knowing how long a course is and how much ground you have already covered makes running a lot easier than when you are ignorant of these facts. Verbal encouragement from people along the route also makes life easier for the runner, as does previous experience with a specific race or forms of exercise. Another good trick is simply to brace yourself: if you expect this to be your hardest race ever, it will probably turn out easier than you thought in the end.

If you have tried every trick in the book and are still unable to maintain your pace, there is one last possibility you can resort to: instead of endeavoring to achieve the time you had set for yourself, you can try to slow your fellow runners down. Not by literally tripping them up, but by using a little psychology. Our brains are very receptive to facial expressions; we try to read them to judge a person's mood. When we unconsciously notice a happy face, it reduces our perception of effort, according to the results of experiments carried out by Marcora. An angry face does exactly the opposite. So if you want to slow your competitors down, wear a T-shirt with a cross face on the back. Just kidding, recreational runners aren't that mean-spirited. We tend to compete only with ourselves.

Research into the psychological side of running has resulted in many new and beneficial insights. It goes without saying that you need an excellent set of physical skills and qualities to become a great athlete. But without the mental equivalent, no runner can ever fulfill their potential.

Mental toughness and psychological skills are much more important to your ability to keep going than was previously thought. And you can learn to persevere too, as long as you get enough practice. Of course, it is your physical fitness that ultimately determines the extent to which you can teach your brain to keep on running. Someone who has difficulty completing a 5K will not be ready to run a marathon after a few weeks of intensive brain training. But the 5K will start to feel a lot easier.

In addition to boosting performance using your psyche, there is another important link between running and the brain. Exercise helps keep the brain and mind healthy. And the effect goes even further: exercise is also a great way of tackling a whole host of mental health problems. But how therapeutic is running, exactly?

TIP

Train your brain to combat fatigue. Go for a run after a long day at work or a bad night's sleep. If you are about to enter a race, avoid all strenuous mental tasks beforehand and set yourself an ambitious but realistic goal, one that will motivate you. If you like to listen to music while running, pick songs whose rhythm will match your stride frequency.

11

RUNNING AS THERAPY FOR THE BRAIN

You can't run and worry at the same time.
—Simon van Woerkom, running therapist

Exercise is good for pretty much everything: digestion, hormones, heart, immune system—you name it. It is also good for the brain, a fact that has become crystal clear over the past fifteen years or so, with the number of studies linking physical exercise to mental well-being continuing to grow rapidly. We now know that exercise has an effect on our cognition and behavior, and that it is linked to a wide range of changes that take place in different parts of the brain. Psychologists and family doctors are increasingly recommending exercise as a way of tackling all kinds of mental health problems—from cognitive decline through old age and mild cognitive disorders, such as memory loss, to diseases like Parkinson's, Alzheimer's, and depression.

Study after study has shown that we need physical exercise to keep our brains in shape and our mental cogwheels in good working order. But is running the best kind of exercise?

If we are to believe the rumors, running scores far higher than walking, cycling, or power training, for example, when it comes to our mental well-being. Running appears to be not only a very effective method for keeping burnout and depression at arm's length, but also for tackling mental problems after they have already manifested themselves. No one doubts that running can make you feel good, and most runners will testify to its healing powers. But can running really work as a kind of mental medicine? And is the effect on the mind so strong that it can even banish depression?

MIND-BLOWING SUBSTANCES

What better place to start than with the most notorious effect running has on the mind: the so-called runner's high, the feeling you get after a long aerobic training session. Science describes it as a feeling of euphoria coupled with diminished anxiety and pain. "Every athlete has heard of it, most seem to believe in it, and many say they have experienced it," according to the *New York Times* reporter Gina Kolata. It is claimed that runner's high literally makes you high when you exercise, as if you've taken hallucinogenic drugs. Not only runners have reported experiencing this high but other endurance athletes, too, like cyclists.

A quick look at a selection of blogs reveals that everyone seems to experience runner's high differently. For some, problems seem to vanish into thin air, while others are a bundle of energy after finishing a long run. Some acquire a feeling of invincibility. Sometimes the feeling comes on after three miles, sometimes after six, or even at the end of a half marathon. Of course, not every runner experiences this high. I for one have never felt it; after a long run, all I ever want to do is plonk myself down on the couch and fall asleep. Extra energy? Not a hope.

Is runner's high a mix of subjective feelings, or is it located somewhere in the brain as a specific kind of activity? Scientists argue about whether or not this euphoric feeling can be traced to a specific location in the brain. They continue to search for the chemicals that are released in the brain as a result of exercise and that are known to alter one's mood. For the past thirty years, the most popular—and most controversial—theory has been that runner's high is sparked off by endorphins. These are the proteins with pain-killing properties that are released in the body during physical exercise, as well as when you experience pain or an orgasm. Endorphins work the same way as opiates; they can ease pain and stimulate a feeling of well-being and happiness, like morphine and heroin do. So it isn't unusual that endorphins are most often identified as the most likely source of runner's high. There is one small problem with this theory, however. When the hypothesis was originally formulated back in the 1980s, it was based primarily on the discovery that there were more endorphins to be found in the blood after a bout of intensive training. But endorphins in the blood are not a reliable indication of what goes on in the brain. In fact, endorphins are not even able to travel from the blood to the brain because of the blood–brain barrier that prevents chemicals like endorphins from entering the brain's cerebrospinal fluid.

For a long time, there was much debate about whether endorphins were released in the brain while running. The German neuroscientist Henning Boecker and his colleagues believe this is exactly what happens. In 2008, they asked ten male athletes to run for two hours and then used medical imaging technology to measure the concentration of naturally occurring opiates in their brains. The difference between the situation before and after was unmistakable: the concentration of opiates was significantly higher in several brain regions after the run, and the runners also felt more euphoric and happier. Not everyone agrees with these findings, however. Other scientists suggest that the difference can be attributed to a placebo effect and that endorphins are not responsible for the perceived feeling of euphoria. Consequently, the link between naturally occurring opiates and runner's high remains a controversial one. Furthermore, endorphins also have to share the attention of scientists with another candidate in the search for the cause of runner's high, one that may have an even stronger claim to the title: naturally occurring cannabis.

NATURALLY OCCURRING CANNABIS

According to some German neurobiologists, the blissful feeling you get from running is the result of a process in the brain that can also be triggered by smoking marijuana. In 2015, Johannes Fuss and his colleagues presented the results of a study in the journal *PNAS* in which the naturally occurring chemical anandamide was identified as a cause of runner's high. Just like endorphins, anandamide also improves your mood, as well as alleviating pain and depression, and it is present in larger quantities in the blood after a training session. But there are also differences between the two. Whereas endorphins work like an opiate, anandamide has the same effect as cannabis. I have never smoked a joint myself, but I have seen the effect it has on others. A person who has just smoked marijuana appears very happy and relaxed—you can see it in their eyes—and it is not difficult to make the link with runner's high.

It gets even better. Unlike endorphins, anandamide can gain access to the brain through the blood and set a "high" in motion. Bingo! Tests on mice in the lab at the University of Heidelberg showed that anandamide was partly responsible. The mice were made to run for five hours in a running wheel. After their workout they exhibited fewer signs of anxious behavior and a higher pain threshold than mice who had not worked out. For example, they stayed standing on a hot plate for longer before jumping off and licking their feet because they could tolerate more pain. To prove that anandamide plays a role in the creation of runner's high, Fuss repeated the experiment. Only this time the mice were injected with substances that blocked either the anandamide or the endorphins. When the endorphins were blocked, not much happened; once again the mice were more relaxed after their workout. But when the anandamide was blocked, the mice remained equally anxious and sensitive to pain no matter how long they ran. It appears, therefore, that running can induce feelings of calm and happiness by activating the body's naturally occurring cannabis. But did the mice feel euphoric as well? Unfortunately, that sensation is too subjective to study in animals, according to Fuss. You can't ask them, after all.

So is naturally occurring cannabis the source of this fantastic feeling? Time to call Professor Erik Scherder, the Netherlands' foremost expert on matters relating to the brain and exercise. What does he think of the cannabis theory? "There are so many different aspects to runner's high that it would be very odd if all the effects could be attributed to one single substance," he tells me on the phone. He explains that exercise also stimulates the production of other naturally occurring chemicals that influence our mood, such as serotonin. "You can never single out one substance as the cause. It always involves a combination." All kinds of chemicals are released in our brain when we take intensive exercise, and the combination is probably different for every individual. Does this difference explain why some people merely feel relaxed during a run while others feel euphoric? We know that anandamide relieves pain, in mice in any case. So if my headache disappears while I'm out running, does that mean I have experienced a form of runner's high? It might not be the most spectacular "kick" in the world, but you won't hear me complaining.

A FIT BRAIN

I often hear people say that they run primarily to "clear the mind." I think what they mean is that this is their way of leaving the day's stress behind them and focusing purely on moving their body and breathing in some fresh air. But a clear mind? I don't know. When I am running, my mind is a whirlwind of thoughts; everything from what I'm going to have for dinner to musings of a more philosophical nature. To me, running feels less like an opportunity to clear the mind and more like a chance to refresh the brain. Running is a reset button, one that allows me to replace work-related worries and stress with fanciful daydreams. And it works.

When I'm stressed about an article that refuses to flow from my pen, a run around the block helps to put everything into perspective. And if I was suffering from a headache too, that usually disappears as well.

A long run can make you feel on top of the world, but it is also a bruising experience for your brain. Intensive exercise takes its toll on your explicit memory, the conscious long-term kind that helps you remember the way home. After running a marathon it can be difficult to access certain information. In a study carried out at Columbia University, runners who had just completed a marathon had trouble remembering words they had drilled into their memory from a list beforehand. The control group, which had taken the test the day before the marathon, was able to recall far more words. The diagnosis? Acute memory loss, arising either from the physical effort involved or from the resulting fatigue. Extreme physical activity is stressful for the body. It releases stress hormones, which, in principle, can be beneficial. They cause you to spring into action when required, for example. These hormones affect different areas of the brain, including the hippocampus, one of the regions responsible for memory and learning. After a marathon the brain functions below par, almost as if it has suffered temporary damage. Fortunately, the stress of a race is very short-lived and your hippocampus is soon back in perfect working order again.

It seems that many runners also suffer a loss of memory with regard to the pain and strain of intensive exercise. After a marathon or other grueling race they are often heard to cry "Never again!"—only to sign up for the next run two months later. It's a bit like giving birth, suggests the Polish psychologist Przemysław Bąbel: women tend to assign the pain of childbirth a lower score a few months after the event compared with immediately afterward. Marathon runners are also often unable to recall how bad the pain was the last time out and so underestimate the difficulties ahead when cheerfully signing up for their next race. Even though some athletes are briefly "of unsound mind" in the aftermath of extreme effort, in the long term physical exercise can be very beneficial for your brain. In fact, moderately intensive exercise is essential if you want to keep your gray matter in good condition. But why, exactly? And which kind of exercise is best?

CRUNCH VERSUS CARDIO

Someone who might be able to enlighten me is Marieke van Heuvelen, a lecturer at UMC Groningen in the Netherlands. She is an expert on the relationship between exercise, old age, and health. The people she studies are not professional athletes or even enthusiastic amateurs, but senior citizens, some of whom are very healthy and some of whom suffer from dementia. Exercise appears to affect the cognitive skills of children and the elderly in particular, she tells me on the phone, as if our brain is more sensitive to exercise at certain times in our life. In the case of children, the brain is still developing, while in the case of older people decay has begun to set in. These seem to be the moments at which exercise can really leave its mark. When I speak with van Heuvelen, she and her colleagues have just completed a study involving three groups of senior citizens suffering from dementia. The first group was instructed to go walking and the second group to walk and take strength training, while the third group, who took no exercise but were exposed to a lot of social interaction, acted as the control group. Their cognitive skills were tested again after nine weeks, or at least the higher brain functions that we need for observing, thinking, remembering, and applying knowledge. The result? Cognition improved the most among those who engaged in a combination of cardio and strength training.

The results for people with dementia are not always clear-cut, says van Heuvelen. One study may show that exercise is beneficial, while the next may reveal very little effect. "In our study we saw some improvement in cognition in the experimental group that took exercise and a deterioration in the inactive control group." Exercise not only delays the decline that comes with old age, but also boosts cognition. However, as soon as you stop taking exercise you lose the effect. "For healthy elderly people, a combination of cardio and strength training also has the greatest effect on the higher control functions of the brain and memory," says van Heuvelen. "And only going for walks is better than only doing strength training."

A combination of endurance and strength training is the best recipe for a healthy brain, a conclusion also reached in a review of the relevant literature in the *British Journal of Sports Medicine*. Combining both forms of training can improve the cognition and memory of people over fifty, regardless of how healthy their brains are. The difference between endurance and strength training is not huge. Endurance sports, such as cycling and jogging, may be a bit more effective when you want to improve the cognitive processes required for a specific goal (e.g., running a race under a certain time). Strength training is an effective way of stimulating the memory (storing and retrieving memories) and the working memory (applying accumulated information in the short term), but is not necessarily more effective than endurance training.

NEW NEURONS

We have a pretty good idea of how the brain reacts, neurobiologically speaking, to physical exercise. Exercise increases the flow of blood to the cerebral cortex, causing blood vessels to expand and brain tissue to take on extra oxygen and nutrients. Sometimes it even stimulates the growth of new cells. Scientists used to believe that in adulthood humans no longer had the capacity to produce new brain cells. Fortunately, this has proved to be untrue, although the rate of production is not as furious as during childhood and puberty.

It may almost sound too good to be true, but exercise can actually drive the production of new neurons in certain parts of the brain, such as the hippocampus. One of the most important players in this process is a protein called BDNF (brain-derived neurotrophic factor). BDNF stimulates the growth of new neurons and the connections between them, while at the same time maintaining existing neurons. This is crucial to keeping neurological diseases like Alzheimer's and Parkinson's at bay. Unfortunately, we produce less and less BDNF the older we get. Compared to healthy individuals, people with dementia and patients suffering from clinical depression have less of it circulating in their blood. The good news is that you can boost the production of this miraculous protein by taking moderately intensive exercise, that is, the kind that causes you to breathe harder and your heart to beat faster.

The link between physical exercise and the health of one's brain has been a proven fact for a couple of decades now. Research carried out in the United States in 2011 revealed that the hippocampus in adults grew when they took cardio training three times a week over the course of a year. The training involved in this case was brisk walking. The effect did not show up in adults who only did stretching exercises. In fact, the hippocampus of those test subjects actually shrank during the study. Not because of the stretching exercises, of course, but simply because they got older. The hippocampus of older people shrinks by 1 to 2 percent each year, with a corresponding increase in the risk of dementia. The test subjects in the endurance training group had more BDNF in their blood and their memory improved too.

In terms of your supply of BDNF, it doesn't really matter whether you prefer endurance training or strength training, as they both help to produce the protein. Exactly how this happens is still a mystery, although research carried out by Bruce Spiegelman, a cellular biologist at Harvard Medical School, suggests that it is proteins outside the brain—those released from the muscle cells during exercise—that drive the production of BDNF inside the brain. BDNF plays a role in both endurance and strength training, but the mechanisms behind them are different, according to Marieke van Heuvelen. "We know from experimental tests on animals that lots of different chemicals are released during each type of training. For example, strength training releases a chemical that combats inflammation, while the production of new blood vessels is stimulated by another chemical that is specific to endurance training."

Taking regular exercise all your life can help keep your brain fit. While it doesn't guarantee that you will never suffer any mental health problems, it certainly reduces the risk. And there is more good news: it is never too late to start. "It's always worth your while starting," says van Heuvelen. "The health of a person who has been inactive all their life will improve faster when they start taking exercise." Even if you only start cycling when you're seventy, the extra shot of BDNF in your brain will help to delay the process of deterioration and sometimes even reverse it.

A single training session can improve both your mood and your cognition, according to an article by two scientists from New York University published in *Brain Plasticity*. However, the effect is temporary and disappears again quite quickly. The neuropsychologist Erik Scherder explains: "For the brain to profit from an active body, you need to take at least thirty minutes of continuous, moderately intensive exercise seven days a week. Cycling, jogging, gardening, or going for a brisk walk are all good forms of exercise." The latest research shows that this is the minimum amount of exercise required to keep the brain in good condition.

BURNOUT

Sport and exercise help to keep the brain healthy. But we have to ask: is running really so effective as a "mental medicine" that it can cure mental health problems? Physical exercise is becoming increasingly accepted in both psychology and psychiatry as a form of therapy. The treatment of depression has long been based on counseling and medication. But now sport and exercise have become part of the equation too.

In scientific literature, running is often identified as a way of helping people who suffer from depression, anxiety, and stress. This kind of training even has its own name: running therapy. Since 2009, the Netherlands has had its own running therapy institute. It was set up by psychiatrist Bram Bakker and physiotherapist Simon van Woerkom, both of whom are qualified "running therapists." They have also written a book on the subject. People with psychiatric problems can sign up for jogging sessions under the supervision of a running therapist. According to their website, "Running therapy can alleviate feelings of depression, but it is actually meant for everyone who wants to improve their mental health and deal with problems like stress, insecurity, anxiety, and depression." So do their claims stand up to scrutiny?

Running appears to be a good idea if you are an exhausted student. However, the last thing most students want to do after cramming all night is put on a pair of running shoes. Wouldn't it only make you more tired? Well, if we are to believe psychologist Juriena de Vries at Radboud University, it is exactly what you should do. Exhaustion and prolonged periods of stress are often early signs of burnout, the last stop for many victims of fatigue. For her doctoral thesis, de Vries asked ninety-nine students to participate in a study. All of them were suffering from severe fatigue, probably because of study-related stress, but they were still capable of studying. None of them took regular exercise. The students were divided into two groups. One group was given six weeks of running therapy, during which they ran twice a week at a relaxed pace under the supervision of a trainer and once a week on their own. The students in the control group did not take any exercise. Each test subject filled out a questionnaire on their mental well-being prior to and immediately after the exercise program. When the program reached its end, the level of fatigue among the joggers had dropped more than that of the inactive test subjects. They also needed less recovery time after a day spent studying, and were able to relax more easily and forget about their work. Furthermore, the joggers indicated that they were able to function much better than the inactive group.

The beneficial effects were still apparent three months after the program ended. "Eighty percent of the students continued taking exercise after the six weeks were up, which may explain this outcome, of course," de Vries tells me on the phone. However, the effects were also apparent among those who had stopped taking exercise immediately after the program. Their condition remained the same, while that of the test subjects who continued to exercise showed some improvement. "It is possible that the effect of running disappears when you stop. That the effect was still noticeable could be attributed to the increased awareness among test subjects during the program of both their study behavior and their fatigue. And, of course, to the fact that they subsequently made changes to their daily routines. All this is pure speculation, however."

In a second study, this time with exhausted workers, running therapy achieved the same success: the test subjects slept better and were better able to perform their work. Running appears, therefore, to be an excellent remedy for study- and work-related fatigue. But can running therapy help at the extreme end of the spectrum—severe burnout—too? De Vries believes it can, despite the fact that someone sitting at home feeling burned out barely has the energy to get up from the couch. The danger is that running could turn out to be another hammer blow for someone who is already in very bad shape. "Exercise may help when the symptoms of exhaustion have become less severe, and then in combination with other forms of therapy."

There is still one more mystery that needs solving: did the students and workers feel better specifically because of the physical exercise they took? Or are there other ingredients in running therapy that deserve the credit instead? It is still not clear why running therapy appears to work. It is probably down to a mix of neurobiological and psychological effects, explains de Vries. "Taking exercise causes more blood to travel to the brain, where chemicals are then released that make you feel good." She also suspects that there is a significant psychological element to the healing powers of exercise. Taking exercise together with others strengthens your social network, for example. Another theory is that success in sport breeds self-confidence, which can result in more confidence at work. And going for a run with a partner can help to banish negative thoughts.

Even just being outside in a green environment is good for one's mood. In 2011, researchers at the University of Exeter reviewed eleven studies of indoor and outdoor exercise and their effect on mental well-being. Taking exercise in natural surroundings inspires feelings of vitality and involvement, they concluded. It reduces tension, confusion, anger, and depression, and boosts your energy levels. And after an outdoor training session we are more likely to return again for another workout.

The best advice de Vries can give to exhausted students and workers is to get regular exercise—moderately intensive exercise, to be precise, because it is important not to overdo it. "If you are already exhausted or stressed out, it is probably not a good idea to subject yourself to an hour-long spinning session because it takes too much energy for you to recover. I recommend taking regular, low-intensity exercise, the kind that still leaves you with enough puff to conduct a conversation. It is also important that the way you exercise allows you to take your mind off work or study, for example by working up a sweat in the great outdoors."

PILLS, COUNSELING, AND RUNNING

Fatigue and chronic stress are troublesome complaints but depression can be extremely debilitating. Can a jog in the park alleviate the symptoms of such a serious condition? There is enough anecdotal evidence that running can help a person to get back on their feet, but it is not supported by scientific literature. The link between physical exercise and depression has been made many times. However, there is very little convincing evidence that exercise works as an antidepressant, primarily due to weak methodology and poorly designed studies. When you look carefully at the small number of well-executed studies that have been carried out, the benefits of running appear less obvious than is often assumed.

Take, for example, the Cochrane review in 2013 of all of the research carried out into depression and physical exercise. The conclusion of the review was that treatment based on physical exercise only has a minor effect on the symptoms of depression in the short term. Exercise is not some kind of cure-all, but it is better than inactivity, a placebo, or meditating. On the other hand, a second meta-analysis carried out by an international research group at the Catholic University of Leuven in 2016 attributed a significantly greater effect to physical exercise. It even went so far as to identify exercise as an evidence-based treatment for depression.

"Studies of the link between exercise and depression point to a moderate effect," says neuropsychologist Erik Scherder. "However, it should also be said that the effect is similar to that achieved through medication." So exercise works just as well as pills, but then without the side-effects of antidepressants, for example, including loss of sexual desire and a dry mouth. In addition to the side effects, another problem is also associated with antidepressants: people often simply don't take their medication. Between 12 and 40 percent of all patients stop taking their pills within the first two months. But doesn't the same apply to exercise? Following an exercise program faithfully for a few months can have its pitfalls, and many patients stop exercising once they have completed the program. Research has shown that the positive effect of exercise on depression disappears if the patient does not continue to exercise.

Juriena de Vries explained to me how it is often difficult for patients to get their health insurance company to cover the costs of running therapy. This may become easier in the future if more evidence can be found to support the benefits of running as a treatment for mental health problems. For the moment, however, that evidence is still lacking, and more high-quality and extensive studies are required first to establish the link between exercise and depression. In the Netherlands, a study was recently completed into the effectiveness of running therapy as a supplementary treatment for patients suffering from chronic depression. Psychiatrist Frank Kruisdijk from the GGz (Dutch Association of Mental

Health Care) carried out the study together with experts from the TNO research institute and the UMC at the Free University in Amsterdam. One group of test subjects ran twice a week for six months as a supplement to their standard treatment with medication and cognitive behavioral therapy. The patients in the control group stuck to the standard course of treatment. The research team measured the improvement in the patients' symptoms after three, six, and twelve months. This study is one of the few that has tried to establish whether running therapy can be used to help patients. "Full-blown clinical depression is a lot different from the kind of depression-related complaints that might make you go and see your doctor," Kruisdijk tells me in an email. "The results were disappointing, to be honest. Many of the test subjects opted out along the way and not one of the clinical patients managed to complete the program. Only a few of the test subjects made it past the three-month mark before also throwing in the towel. I haven't reached the point yet of advising against exercise as a form of treatment for the most severely ill patients, but I'm not that far off either." Over half of the forty-six test subjects stopped halfway, and only at the three-month mark were there enough patients to compare the groups with each other. The results showed that the symptoms had improved slightly in both groups, regardless of whether they ran or not, although the patients in the running group did appear to be fitter.

After all the success stories about how running can help combat fatigue and stress, this is a bit of a disappointment. As de Vries already mentioned, it's hard enough just to get out of bed when you are suffering from severe burnout or

depression, never mind go for a run in the park. Frank Kruisdijk thinks that, for the moment, we should reserve running
therapy for highly motivated patients suffering from light to
moderate depression; people who are still able to work, take
regular exercise or run regularly, and are suffering from burnout. "In the case of chronic depression, a multidisciplinary
lifestyle intervention that focuses on aspects like quitting
smoking and eating healthily appears more suitable. The
preventive powers of exercise when it comes to combatting
the onset of depression and the likelihood of curing clinical
depression by running are two different things entirely." De
Vries does see a role for the so-called stepped care approach
to treating people with light to moderate depression. This
involves starting with a less intensive form of therapy with
few side effects, such as running therapy. If that doesn't
help, then the decision can be taken to move on to other
more intensive forms.

Neuropsychologist Erik Scherder also sees a role for running in treating depression, albeit a limited one. "If you were
to ask me whether I think running therapy helps, I would
say yes. But only as a supplementary form of treatment in
addition to medication and cognitive behavioral therapy
and not as a stand-alone treatment."

ADDICTED TO EXERCISE

Can being physically active also have a negative effect on
your mental well-being? I often get the heebie-jeebies when
people say to me that to them running is "a way of life." It
sounds a bit too obsessive for my liking, as if your whole life
revolves around running as far and as fast you can. It might

have something do with the tale of Scarlett Thomas, the British sport junkie who told her story to the *Guardian*.

Thomas was addicted to drugs until the kick she got from exercise proved better than the one she got from cocaine. Her perfect way of spending a day was to go for a Pilates session at the gym, nip home for a cup of tea, go out again to play tennis for two or three hours, back home for a quick shower and a bowl of complex carbohydrates, and then out for the evening to play a tennis competition. It all started around the age of thirty. She became a fitness instructor and took an online course in nutrition. Exercise felt so good that she just couldn't stop. Thomas compiled statistics and kept charts of her weight and the number of steps she took each day. Thanks to her Fitbit, an electronic wristband, she always knew how active she was at any given moment. She followed the Fitbit accounts of her friends as well so that she could cover more miles each day than they did. Her wristband also kept track of her calorie intake. One day Thomas realized that her behavior was becoming obsessive, but she didn't think it would become a problem; if anything, it was a healthy obsession. There is a great irony to her story. Running can be an effective medicine against mental health problems like chronic stress and exhaustion, but when you take it too far it becomes the exact opposite: an unhealthy activity that causes stress. In 1997, the British psychologist Mark Griffiths described the experiences of an obsessive athlete in the journal *Addiction Research*. He called her Joanna—not her real name. Joanna was twenty-five years old. Despite a recurring injury, she stuck rigidly to her extremely demanding training schedule until eventually her boyfriend got so fed up with her that he ended their relationship. Although she did not describe herself as addicted, she did realize she had a problem.

There are plenty of people in the world like Joanna. In 2011, as part of her master's degree at Radboud University, Sabine Janssen undertook a study of sports addiction, or sports dependency as she prefers to call it herself. Using a questionnaire, she surveyed the exercise habits and feelings of 1,700 students at the Radboud Sports Center. The questions included the number of hours they dedicated each week to exercise and their sources of motivation. She concluded that 1.5 percent of the athletes were addicted to exercise; these obsessive fanatics were busy with sporting activities 7.5 hours a week, primarily with the aim of combatting irritation and stress. Less was not an option for them. More than a third of those surveyed showed some symptoms of exercise addiction, such as dissatisfaction with their body, a desire for more control over their own lives, and a dependency on the calming effects of exercise. An excessive amount of exercise is especially bad for your mental health. Janssen found that exercise addicts will always choose their exercise regime over other duties and responsibilities. Men spend less time with their family, while women sacrifice their career in favor of their training schedule.

Exercise addiction is still a hot potato in the scientific community. In the fifth edition of the *Diagnostic and Statistical Manual of Mental Disorder*, the standard manual for psychiatric disorders, exercise addiction is not (yet) recognized as an addictive form of behavior (neither are excessive shopping, eating, or gaming). Before you can speak of an addiction to exercise, you have to be sure that the behavior is not the result of some other disorder. For example, over 40 percent of people with an eating disorder, like anorexia nervosa, also take excessive amounts of exercise, with the latter problem often being the result of the former.

SELF-CONFIDENCE, STRESS, AND A LEANER BODY

I train between seven and eight hours a week and have no intention of cutting down. I enjoy a bit of activity after a long day spent working at the computer. I get itchy feet when I haven't pulled on my running shoes for two days and I plan my training schedule very carefully. And yet it is not an obsession. Apart from workout junkie, I'm also a dedicated couch potato. I can spend whole evenings stretched out on the couch or days lying on the beach without feeling any guilt or a compulsion to move. Running is not the most important thing in my life, but it's not that far off either.

How bad is it, actually, to be addicted to physical exercise? Surely it is better to be obsessed with sport than with drugs. Clinical psychologist and specialist in behavioral addiction Marilyn Freimuth at the Fielding Graduate University in California does not agree entirely with this hypothesis, she tells me in an email. "Some people exercise so often that it has a negative effect on their daily lives." These effects can be physical in nature (injuries), but the damage can be even greater in the interpersonal sphere. Exercise addicts often argue with their partner about the amount of time they spend in the gym, and some relationships collapse under the strain. "Despite the problems that arise from their behavior, many find it very difficult to stop or cut down. These two outcomes—a negative effect on daily life and difficulty stopping—are important characteristics of addiction."

We are not all equally susceptible to becoming addicted to exercise. However, sports science students appear to run a higher risk than most; at 7 percent their risk of addiction is twice as high as that for members of the general public who are regular visitors to the gym. This may be because sports science students have a special interest in sport or because a sports college education attracts types who are more prone to addiction.

It also depends on why you take lots of exercise. People who run primarily with the aim of improving their times are unlikely to encounter any problems. Those of us who run to alleviate stress or anxiety, however, have a greater risk of becoming addicted, as do people who run in pursuit of a slimmer body or those whose identity is heavily dependent on being an athlete. Freimuth can tell when someone has entered the danger zone: "Visiting the gym more often and for increasingly longer periods of time is a sign that you are heading down the road to addiction." There is nothing wrong with running to build your self-confidence, as long as you don't lose control. Luckily, the vast majority of runners are not addicted, however ambitious and dedicated they might be. When you get injured or sick you should just stop. Period. That way running can remain a healthy habit.

TIP

Are you struggling with study- or work-related fatigue? Try to get regular exercise. Moderately intensive exercise, such as cycling or jogging, works best. It's okay to run to reduce your stress levels, but make sure your sense of well-being doesn't become dependent on exercise.

EPILOGUE: DO YOUR OWN SCIENCE

Running is real and relatively simple . . . but it ain't easy.
—Mark Will-Weber, former senior editor at *Runner's World* magazine

When I began my search for the science behind running, I had fully expected to find answers to most of my questions. And in many cases I did. Hydrating too much while running can be dangerous, no doubt about that. The less well-trained you are, the more you will benefit from beetroot juice, all the evidence points in that direction. Without a high VO_2max you will never be an outstanding marathon runner, simple as that. And too little sleep can have a disastrous effect on performance. Medical and sports sciences sometimes come up with answers that you just can't argue with. But sometimes those answers are less than conclusive, especially when they concern matters like running technique and training routines. In this book I have often had to reach the conclusion that there is simply "no evidence" to support a particular claim. Not for stretching, not for compression socks, not for a certain type of shoe. Of course, it would be easy to dismiss many of these claims as nonsense, but that wouldn't be prudent either. Is there no evidence because no evidence exists? Or is there no evidence because the right studies haven't been carried out yet?

Take the way you land your foot, for example. There is no scientific reason to suggest that everyone should start landing on their forefoot instead of their heel; in fact, instead of preventing injury, this running technique can spark a whole new set of complaints. However, there is some evidence that switching from heel to forefoot can benefit certain runners. But who are those runners? And under which conditions should they make that switch? Reasonable questions to which scientific research has yet to provide satisfactory answers. What should you do in this situation? You could choose, like I did, to switch to landing on your forefoot in response to a specific problem. After all, it's only by trying something that you can find out if it works for you. You do need to remember, however, that in the first instance switching from heel to forefoot actually increases the risk of injury.

One thing you always need to keep in the back of your mind is that the results of studies using test subjects are always stated in terms of "averages." Scientists conduct their studies on the "average" runner, while in real life we are all individuals, each with our own unique situation. Imagine that a study shows that taking caffeine on board before a race can improve your performance by an average of 1 percent. Should we all immediately start stocking up on coffee? It is highly likely that some of the test subjects in such a study exhibit a relatively insignificant reaction to caffeine (e.g., for genetic reasons), while others end up running not 1 but 2 percent faster. There may even be a test subject or two whose running suffers (maybe because of stomach cramps) as a result of consuming caffeine. Which of these people are you? Most studies don't tell us this kind of stuff.

Fortunately, we can also carry out a little scientific research of our own on our body. Results from scientific literature give us a good idea where to start. For example, if caffeine has been shown to have a positive effect during a 10K run, we can test the theory ourselves. Just get some caffeine into your system before your next race and see what happens. Chances are it will boost your performance. If, however, you happen to be the kind of person who gets stomach cramps instead, that's a bummer, but at least now you know. Trust your own observations and analyze your own experiences. As the sports physician Backx once said: "It's not just about the evidence, but also about the experience."

With regard to running shoes, you can adopt the "do your own science" approach as well. The biomechanical engineer Benno Nigg at the University of Calgary came up with a new paradigm that dovetails nicely with this approach. In his theory, it is not foot type that determines your choice of running shoe but your "comfort filter." He describes comfort as a feeling that matches your "preferred movement path": the unique way your body moves naturally in order to conserve as much energy as possible. Nigg believes that our body naturally adopts a certain style of running. In an experiment he conducted a few years ago, most of the runners ran the very same way when they switched from one type of shoe to another.

Instead of correcting the way you run, running shoes should actually support your preferred movement path. According to the theory, your shoes should not be allowed to push you off that path because your muscles will only end up having to work harder, which can result in injury. The best way to find shoes that match your own style of running is to follow your own comfort filter. In other words, trust your intuition. Based on the most recent scientific information, assistants in footwear stores should focus on finding the most comfortable kind of shoe for your feet, not one that tries to "correct" the way you run. In an ideal world, you would be allowed to take your new shoes home for a week to see if they are still comfortable when you train in them.

I'd like to let you in on a little secret. Despite the fact that I am usually only ever swayed by hard evidence, my compression socks have somehow found their way back into my socks drawer. They were rescued from the back of my closet prior to a 14K race on a course with unpaved tracks and hills that reached a height of 700 feet. A hell of a challenge for your calves and thighs (especially if you are more used to the flatlands of the Netherlands). The research into the effect of compression socks popped into my mind: there is no experimental evidence that they prevent injury, I recalled. But still I decided to wrap my calves up tightly for some extra support. Just in case. The race was tough, but it went smoothly and better than expected: I crossed the line in a time of 1:09. It's hard to say, of course, how I would have run without the socks; probably the same. But that day it felt good to pull them on, like I was preparing myself well. I assigned a certain significance to the socks and they gave me that extra bit of confidence I needed to tackle the hills. "If you feel that something will help, then it will help," as the physiotherapist Maarten van der Worp puts it. The exercise physiologist Samuele Marcora said something similar: "The fact that something is perceived does not make it any less real or forceful."

"Science is nothing but perception," the Greek philosopher Plato is reputed to have said, but I wouldn't go that far. Sometimes the evidence is so overwhelming that you can't change the facts no matter how much you argue. "The good thing about science is: it's true whether you believe it or not," the renowned astrophysicist Neil deGrasse Tyson tweeted in 2013. But we all know how important personal convictions and expectations can be when it comes to our own body: if you think that something is going to work, it often does. If you are convinced that you will run faster on a muesli bar than on a peanut butter and jelly sandwich, then you probably will, and not because the muesli bar contains some kind of secret fuel. The above should not, of course, be interpreted as a license to start using pseudoscientific gimmicks, like wrapping yourself in expensive kinesiology tape to ward off injury, wearing balance bracelets to keep body and mind on an even keel, or smearing your body with amino acid cream to aid muscle recovery. The same applies to all kinds of vague dietary supplements. At best they are nothing more than a waste of money, according to the medical biotechnician Stephan Peters in an article in the Dutch newspaper *de Volkskrant*.

I will never buy myself a pair of anti-pronation shoes again. Both science and personal experience have made sure of that. Although it remains a process of trial and error, I believe I am slowly figuring out what works best for me. Beetroot juice is not on my list. No matter how promising the results might be in terms of enhancing performance, I won't be letting that awful stuff past my lips ever again. Sometimes running is simply a matter of taste.

ACKNOWLEDGMENTS

It took me almost two years to write the original Dutch version of this book. Fortunately, Ward Weistra, my partner and fellow biologist, never doubted for one moment that I would complete the task. He afforded me the time and space I needed to work on the project and, more importantly, encouraged me to think big and be more assertive. He even offered to work as my PR manager so that he could help promote my work. Many of the articles I wrote about running and exercise for publications like *Runner's World* have been recycled in this book.

My gratitude for this translation goes out to acquisitions editor Anne-Marie Bono at the MIT Press for her faith in this book. She was enthusiastic about setting the arrangements for *Running Smart* in motion. Manuscript editor Judy Feldmann did an extremely thorough edit and was most helpful in improving the book's flow. Danny Guinan did an amazing job translating the book. The English text turned out more beautiful than the original. I am also very grateful for the help of the many experts who offered their time and advice so graciously to me. Thanks to their careful proofreading, the number of errors has (hopefully) been kept to a minimum. Their expertise was the source of many valuable insights and suggestions, which only helped to reinforce the scientific character of the book. So, in no particular order, thanks to: Steef Bredeweg, Ron Diercks, Hidde Haisma, Maarten van der Worp, Marienke van Middelkoop, Frank Backx, Dionne Noordhof, Jenny Hofstede, Rinze ter Steege, Thijs Eijsvogels, Erik Hulzebos, Marije Elferink-Gemser, Vana Hutter, Benedicte Vanwanseele, Ellen Maas, Juriena de Vries, and Marieke van Heuvelen.

I also want to thank my parents, Gerard and Janet, who support me in everything I do. It is from them that I learned the importance of following your heart. And that is exactly what I have done by combining two of my favorite pastimes—running and science—in this book. I have never felt under any external pressure to work harder or perform better. Any pressure I did feel was all of my own making. Which brings me to another group of people I would like to thank: the team at the Revalide rehabilitation center. In late 2015 I was forced to stop working for a few months because of a persistent case of RSI. The people at Revalide helped me to get back in the saddle again step by step. I am sure that without their care, both physical and mental, I would never have been able to pick up a pen and write again, never mind pursue a career as a science journalist.

Finally, many thanks to my colleagues at the AV Phoenix running club for your friendship and the good times we share. Not to mention the hilarious anecdotes that made it into this book. My passion for running only truly began to burn when I started running with you!

BIBLIOGRAPHY

1 THE RISE OF THE LONG-DISTANCE RUNNER

Alexander, R. M. "Energy-Saving Mechanisms in Walking and Running." *Journal of Experimental Biology* 160 (1991): 55–69.

Balke, B., and C. Snow. "Anthropological and Physiological Observations on Tarahumara Endurance Runners." *American Journal of Physical Anthropology* 23, no. 3 (1965): 293–301.

Bramble, D. M., and D. E. Lieberman. "Endurance Running and the Evolution of Homo." *Nature* 432, no 7015 (2004): 345–352.

DeSilva, J. M., and Z. J. Throckmorton. "Lucy's Flat Feet: The Relationship between the Ankle and Rearfoot Arching in Early Hominins." *PLOS ONE* 5, no. 12 (2010: e14432.

DeSilva, J. M., K. G. Holt, S. E. Churchill, K. J. Carlson, C. S. Walker, B. Zipfel, and L. R. Berger. "The Lower Limb and Mechanics of Walking in *Australopithecus sediba*." *Science* 340, no. 6129 (2013): 1232999.

Harcourt-Smith, W. E. H., and L. C. Aiello. "Fossils, Feet and the Evolution of Human Bipedal Locomotion." *Journal of Anatomy* 204, no. 5 (2004): 403–416.

IAAF. "London 2012 Paralympic Games. Oscar Pistorius' Blades: An Annotated Graphic." *Engineering & Technology Magazine*, September 19, 2012.

IAAF. "Oscar Pistorius: Independent Scientific Study Concludes That Cheetah Prosthetics Offer Clear Mechanical Advantages." January 14, 2008.

Liebenberg, L. "Persistence Hunting by Modern Hunter-Gatherers." *Current Anthropology* 47, no. 6 (2006): 1017–1025.

Lieberman, D. E. and D. M. Bramble. "The Evolution of Marathon Running Capabilities." *Sports Medicine* 37, nos. 4–5 (2007): 288–290.

Lieberman, D. E., D. M. Bramble, D. A. Raichlen, and J. J. Shea. "The Evolution of Endurance Running and the Tyranny of Ethnography: A Reply to Pickering and Bunn (2007)." *Journal of Human Evolution* 53, no. 4 (2007): 439–442.

Lieberman, D. E., D. A. Raichlen, H. Pontzer, D. M. Bramble, and E. Cutright-Smith. "The Human Gluteus Maximus and Its Role in Running." *Journal of Experimental Biology* 209, no. 11 (2006): 2143–2155.

Lovejoy, C. O., G. Suwa, L. Spurlock, B. Asfaw, and T. D. White. "The Pelvis and Femur of *Ardipithecus ramidus*: The Emergence of Upright Walking." *Science* 326, no. 5949 (2009): 71e1–6.

McDougall, C. *Born to Run: The Hidden Tribe, the Ultra-Runners and the Greatest Race the World Has Never Seen.* London: Profile Books, 2009.

Pickering, T. R., and H. T. Bunn. "The Endurance Running Hypothesis and Hunting and Scavenging in Savanna-Woodlands." *Journal of Human Evolution* 53, no. 4 (2007): 434–438.

Poelman, Y. *De natuur als uitvinder: Miljarden jaren aan innovatie gratis beschikbaar.* Amsterdam: Uitgeverij Carrera, 2015.

Scheerder, J., K. Breedveld, and J. Borgers, eds. *Running across Europe: The Rise and Size of One of the Largest Sport Markets.* London: Palgrave Macmillan, 2015.

Sports Marketing Surveys Inc. "Running Population UK." August 12, 2014. https://www.sports-insight.co.uk/market-analysis/The-UKs-running -population-has-reached-10.5million-runners-This-means-runn.

Roy Morgan Research. "Over 1.1 Million Australians Swim, Cycle and Run—but Only a Few Are True Triathletes." March 1, 2016. http:// www.roymorgan.com/findings/6700-australians-who-swim-run-and -cycle-and-do-triathlons-december-2015-201603010006.

Weyand, P. G., M. W. Bundle, C. P. McGowan, A. Grabowski, M. B. Brown, R. Kram, and H. Herr. "The Fastest Runner on Artificial Legs: Different Limbs, Similar Function?" *Journal of Applied Physiology* 107, no. 3 (2009): 903–911.

2 THE USEFULNESS OF RUNNING SHOES

Chambon, N., V. Sevrez, Q. H. Ly, N. Guégen, E. Berton, and G. Rao. "Aging of Running Shoes and Its Effect on Mechanical and Biomechanical Variables: Implications for Runners." *Journal of Sports Sciences* 32, no. 11 (2014): 1013–1022.

Keller, T. S., A. M. Weisberger, J. L. Ray, S. S. Hasan, R. G. Shiavi, and D. M. Spengler. "Relationship between Vertical Ground Reaction Force and Speed during Walking, Slow Jogging, and Running." *Clinical Biomechanics* 11, no. 5 (1996): 253–259.

Lieberman, D. E., M. Venkadesan, W. A. Werbel, A. I. Daoud, S. D'Andrea, I. S. Davis, et al. "Foot Strike Patterns and Collision Forces in Habitually Barefoot versus Shod Runners." *Nature* 463, no. 7280 (2010): 531–535.

Malisoux, L., N. Chambon, N. Delattre, N. Guéguen, A. Urhausen, and D. Theisen. "Injury Risk in Runners Using Standard or Motion Control Shoes: A Randomised Controlled Trial with Participant and Assessor Blinding." *British Journal of Sports Medicine* 50, no 8 (2016): 1–7.

Marti, B., J. P. Vader, C. E. Minder, and T. Abelin. "On the Epidemiology of Running Injuries: The 1984 Bern Grand-Prix Study." *American Journal of Sports Medicine* 16, no. 3 (1988): 285–294.

Nielsen, R. O., I. Buist, E. T. Parner, E. A. Nohr, H. Sørensen, M. Lind, and S. Rasmussen. "Foot Pronation Is Not Associated with Increased Injury Risk in Novice Runners Wearing a Neutral Shoe: A 1-Year Prospective Cohort Study." *British Journal of Sports Medicine* 48, no. 6 (2014): 440–447.

Nigg, B. M. "The Role of Impact Forces and Foot Pronation: A New Paradigm." *Clinical Journal of Sports Medicine* 11, no. 1 (2009): 2–9.

Nigg, B. M., J. Baltich, S. Hoerzer, and H. Enders. "Running Shoes and Running Injuries: Mythbusting and a Proposal for Two New

Paradigms: 'Preferred Movement Path' and 'Comfort Filter.'" *British Journal of Sports Medicine* 49, no. 20 (2015): 1290–1294.

Richards, C. E., P. J. Magin, and R. Callister. "Is Your Prescription of Distance Running Shoes Evidence-Based?" *British Journal of Sports Medicine* 43, no. 3 (2008): 159–162.

Van der Worp, H., J. W. Vrielinnk, and S. W. Bredeweg. "Do Runners Who Suffer Injuries Have Higher Vertical Ground Reaction Forces Than Those Who Remain Injury-Free? A Systematic Review and Meta-Analysis." *British Journal of Sports Medicine* 50, no. 8 (2016): 450–457.

Van Gent, R. N. "Incidence and Determinants of Lower Extremity Running Injuries in Long Distance Runners: A Systematic Review." *British Journal of Sports Medicine* 40, no. 8 (2007): 469–480.

Stam Christine, and Huib Valkenberg. *Sport Injuries in the Netherlands —2018.* September 2019. https://www.veiligheid.nl/sportblessures /feiten-cijfers.

Zadpoor, A. A., and A. A. Nikooyan. "The Relationship between Lower-Extremity Stress Fractures and the Ground Reaction Force: A Systematic Review." *Clinical Biomechanics* 26, no. 1 (2011): 23–28.

3 BACK TO BARE FEET

Altman, A. R., and I. S. Davis. "Barefoot Running: Biomechanics and Implications for Running Injuries." *Current Sports Medicine Reports* 11, no. 5 (2012): 244–250.

Bergstra, S. A., B. Kluitenberg, R. Dekker, S. W. Bredeweg, K. Postema, E. R. Van den heuvel, et al. "Running with a Minimalist Shoe Increases Plantar Pressure in the Forefoot Region of Healthy Female Runners." *Journal of Science and Medicine in Sport* 18, no. 4 (2015): 463–468.

Franz, J. R. "Metabolic Cost of Running Barefoot versus Shod: Is Lighter Better?" *Medicine and Science in Sports and Exercise* 44, no. 8 (2012): 1519–1525.

Goss, D. L., and M. T. Goss. "Relationships among Self-Reported Shoe Type, Footstrike Pattern, and Injury Incidence." *U.S. Army Medical Department Journal* (October 2012): 25–30.

Hall, J., C. Barton, P. R. Jones, and D. Morrissey. "The Biomechanical Differences between Barefoot and Shod Distance Running: A Systematic Review and Preliminary Meta-Analysis." *Sports Medicine* 43, no. 12 (2013): 1335–1353.

Hamill, J., and A. H. Gruber. "Is Changing Footstrike Pattern Beneficial to Runners?" *Journal of Sport and Health Science* 6, no. 2 (2017): 146–153.

Hannigan J. J., and C. D. Pollard. "A 6-Week Transition to Maximal Running Shoes Does Not Change Running Biomechanics." *American Journal of Sports Medicine* 47, no. 4 (2019): 968–973.

Hatala, K. G. "Variation in Foot Strike Patterns during Running among Habitually Barefoot Populations." *PLOS ONE* 8, no. 1 (2013): 1335–1353.

Larson, P., R. Y. L. Wong, T. K. W. Chung, R. T. Choi, W. W. Y. Leung, and D. H. Y. Shek. "Foot Strike Patterns of Recreational and Sub-Elite Runners in a Long-Distance Road Race." *Journal of Sports Sciences* 30, no. 12 (2012): 1275–1283.

McKeon, P. O., J. Hertel, D. Bramble, and I. Davis. "The Foot Core System: A New Paradigm for Understanding Intrinsic Foot Muscle Function." *British Journal of Sports Medicine* 49, no. 290 (2015): 1–9.

Middlebrook, H. "Do Mega-Cushioned Shoes Increase or Reduce Injuries? It's Complicated." *Runner's World*, April 30, 2019 (online).

Murphy, S. "Heel Striking. Is It Really the Enemy of Good Running Form?" *Guardian*, October 9, 2014.

Pollard, C. D. "Influence of Maximal Running Shoes on Biomechanics Before and After a 5K Run." *Orthopaedic Journal of Sports Medicine* 6, no. 6 (2018).

Rice, H. M., S. T. Jamison, and I. S. Davis. "Footwear Matters: Influence of Footwear and Foot Strike on Load Rates during Running." *Medicine & Science in Sports & Exercise* 48, no. 12 (2016): 2462–2468.

Robbins, S. E., and A. M. Hanna. "Running-Related Injury Prevention through Barefoot Adaptations." *Medicine & Science in Sports & Exercise* 19, no. 2 (1987): 148–156.

Roth, J., J. Neumann, and M. Tao. "Orthopaedic Perspective on Bare-foot and Minimalist Running." *Journal of the American Academy of Orthopaedic Surgeons* 24, no. 3 (2016): 180–187.

Tam, N., J. L. A. Wilson, T. D. Noakes, and R. Tucker. "Barefoot Running: An Evaluation of Current Hypothesis, Future Research and Clinical Applications." *British Journal of Sports Medicine* 48, no. 5 (2014): 349–355.

Thomas, E. "Vibram, 'Barefoot Running Shoe' Company, Settles Multi-Million Dollar Lawsuit." *Huffington Post*, May 13, 2014.

Vincent, H. K., and K. R. Vincent. *ACSM Information on Selecting Running Shoes*. American College of Sports Medicine, Consumer Information Committee, 2014.

4 BUILT TO RUN

Anderson, T. "Biomechanics and Running Economy." *Sports Medicine* 22, no. 2 (1996): 76–89.

Ash, G. I., R. A. Scott, M. Deason, T. A. Dawson, B. Wolde, Z. Bekele, et al. "No Association between ACE Gene Variation and Endurance Athlete Status in Ethiopians." *Medicine and Science in Sports and Exercise* 43, no. 4 (2011): 590–597.

Brouwer, J. "Deze bedrijven beloven een beter leven door je je DNA te laten testen: Alleen maken ze die beloften niet waar." *De Correspondent*, July 8, 2016.

Brown, N., J. White, A. Brasher, and J. Scurr. "An Investigation into Breast Support and Sports Bra Use in Female Runners of the 2012 London Marathon." *Journal of Sports Sciences* 32, no. 9 (2014): 801–809.

De Moor, M. H. M., T. D. Spector, L. F. Cherkas, and M. Falchi. "Genome-wide Linkage Scan for Athlete Status in 700 British Female DZ Twin Pairs." *Twin Research and Human Genetics* 10, no. 6 (2007): 812–820.

Katz, B. "The Incredible Story of Bobbi Gibb, the First Woman to Run the Boston Marathon." *New York Times*, April 20, 2015.

Ma, F., Y. Yang, X. Li, F. Zhou, C. Gao, M. Li, and L Gao. "The Association of Sport Performance with ACE and ACTN3 Genetic

Polymorphisms: A Systematic Review and Meta-Analysis." *PLOS ONE* 8, no. 1 (2013): e54685.

MacArthur, D. G., J. T. Seto, J. M. Raftery, K. G. Quinlan, G. A. Huttley, J. W. Hook, et al. "Loss of ACTN3 Gene Function Alters Mouse Muscle Metabolism and Shows Evidence of Positive Selection in Humans." *Nature Genetics* 39, no. 10 (2007): 1261–1265.

Montgomery, H. E. "Human Gene for Physical Performance." *Nature* 393, no. 6682 (1998): 221–222.

Noehren, B., J. Hamill, and I. Davis. "Prospective Evidence for a Hip Etiology in Patellofemoral Pain." *Medicine and Science in Sports and Exercise* 45, no. 6 (2013):1120–1124.

Saunders, P. U., D. B. Pyne, R. D. Telford, and J. A. Hawley. "Factors Affecting Running Economy in Trained Distance Runners." *Sports Medicine* 34, no. 7 (2004): 465–485.

Scurr, J. C., J. L. White, and W. Hedger. "Supported and Unsupported Breast Displacement in Three Dimensions across Treadmill Activity Levels." *Journal of Sports Sciences* 29, no. 1 (2011): 55–61.

Scurr, J. C., N. Brown, J. Smith, A. Brasher, D. Risius, and A. Marczyk. "The Influence of the Breast on Sport and Exercise Participation in School Girls in the United Kingdom." *Journal of Adolescent Health* 58, no. 2 (2016): 167–173.

Stokvis, R. "Hardlopende vrouwen in de openbare ruimte." *Sociologie* 2, no. 3 (2006): 249–264.

Timmons, J. A., S. Knudsen, T. Rankinen, L. G. Koch, M. Sarzynski, T. Jensen, et al. "Using Molecular Classification to Predict Gains in Maximal Aerobic Capacity Following Endurance Exercise Training in Humans." *Journal of Applied Physiology* 108, no. 6 (2010): 1487–1496.

Van Rossum, M. "De mythe van het sportlichaam." *De Volkskrant*, February 28, 2004.

Williams, A. G., M. P. Rayson, M. Jubb, M. World, D. R. Woods, M. Hayward, et al. "The ACE Gene and Muscle Performance." *Nature* 403, no. 6770 (2000): 614.

Wilson, J. M., J. P. Loenneke, E. Jo, G. J. Wilson, M. C. Zourdos, and J.-S. Kim. "The Effects of Endurance, Strength, and Power Training on Muscle Fiber Type Shifting." *Journal of Strength & Conditioning Research* 26, no. 6 (2012): 1724–1729.

Yang, N., D. G. MacArthur, J. P. Gulbin, A. G. Hahn, A. H. Beggs, S. Easteal, and K. North. "ACTN3 Genotype Is Associated with Human Elite Athletic Performance." *American Journal of Human Genetics* 73, no. 3 (2003): 627–631.

5 TRAINING LOAD AND LOAD CAPACITY

Alentorn-Geli, E., K. Samuelsson, V. Musahl, C. L. Green, M. Bhandari, and J. Karlsson. "The Association of Recreational and Competitive Running with Hip and Knee Osteoarthritis: A Systematic Review and Meta-Analysis." *Journal of Orthopaedic and Sports Physical Therapy* 47, no. 6 (2017): 373–390.

Ali, A., M. P. Caine, and B. G. Snow. "Graduated Compression Stockings: Physiological and Perceptual Responses during and after Exercise." *Journal of Sports Sciences* 25, no. 4 (2007): 413–419.

Baxter, C., L. R. McNaughton, A. Sparks, L. Norton, and D. Bentley. "Impact of Stretching on the Performance and Injury Risk of Long-Distance Runners." *Research in Sports Medicine* 25, no. 1 (2017): 78–90.

Behm, D. G., A. J. Blazevich, A. D. Kay, and M. McHugh. "Acute Effects of Muscle Stretching on Physical Performance, Range of Motion, and Injury Incidence in Healthy Active Individuals: A Systematic Review." *Applied Physiology, Nutrition, and Metabolism* 41, no. 1 (2016): 1–11.

Bovenschen, J., M. te Booij, and C. J. M. van der Vleuten. "Graduated Compression Stockings for Runners: Friend, Foe or Fake?" *Journal of Athletic Training* 48, no. 2 (2013): 226–232.

Butler, R. J., H. P. Crowell III, and I. McClay Davis. "Lower Extremity Stiffness: Implications for Performance and Injury." *Clinical Biomechanics* 18, no. 6 (2003): 511–517.

Cheatham, S. W., M. J. Kolber, M. Cain, and M. Lee. "The Effects of Self-Myofascial Release Using a Foam Roll or Roller Massager on Joint Range of Motion, Muscle Recovery, and Performance: A Systematic

Review." *International Journal of Sports Physical Therapy* 10, no. 6 (2015): 827–838.

De Ruiter, C. J., P. W. Verdijk, W. Werker, M. J. Zuidema, and A. de Haanl. "Stride Frequency in Relation to Oxygen Consumption in Experienced and Novice Runners." *European Journal of Sport Science* 14, no. 3 (2014): 251–258.

Dicharry, J. *Anatomy for Runners.* New York: Skyhorse Publishing, 2012.

e Lima, K. M. M., S. Carneiro, D. de S. Alves, C. Peixinho, and L. de Oliveira. "Assessment of Muscle Architecture of the Biceps Femoris and Vastus Lateralis by Ultrasound after a Chronic Stretching Program." *Clinical Journal of Sport Medicine* 25, no. 1 (2015): 55–60.

Ferris, D. P., K. Liang, and C. T. Farley. "Runners Adjust Leg Stiffness for Their 1st Step on a New Running Surface." *Journal of Biomechanics* 32, no. 8 (1999): 787–794.

Freitas, S. R., B. Mendes, G. Le Sant, R. J. Andrade, A. Nordez, and Z. Milanovic. "Can Chronic Stretching Change the Muscle-Tendon Mechanical Properties? A Review." *Scandinavian Journal of Medicine & Science in Sports* 28, no. 3 (2018): 794–806.

Freiwald, J. "Foam-Rolling in Sport and Therapy. Potential Benefits and Risks: Part 2: Positive and Adverse Effects on Athletic Performance." *Sports Orthopaedics and Traumatology* 32, no. 3 (2016): 267–2752.

Fu, W., Y. Fang, D. M. S. Liu, L. Wang, S. Ren, and Y. Liu. "Surface Effects on in-Shoe Plantar Pressure and Tibial Impact during Running." *Journal of Sport and Health Science* 4, no. 4 (2015): 384–390.

Hill, J., G. Howatson, K. van Someren, J. Leeder, and C. Pedlar. "Compression Garments and Recovery from Exercise-Induced Muscle Damage: A Meta-Analysis." *British Journal of Sports Medicine* 48, no. 18 (2014): 1340–1346.

Hreljac, A. "Impact and Overuse Injuries in Runners." *Medicine & Science in Sports & Exercise* 36, no. 5 (2004): 845–849.

Hunter, I. "Self-Optimization of Stride Length among Experienced and Inexperienced Runners." *International Journal of Exercise Science* 10, no. 3 (2017): 446–453.

Kemmler, W., S. von Stengel, C. Köckritz, J. Mayhew, A. Wassermann, and J. Zapf. "Effect of Compression Stockings on Running Performance in Male Runners." *Journal of Strength & Conditioning Research* 23, no. 1 (2009): 101–105.

Leetun, D. T., M. L. Ireland, J. D. Willson, B. T. Ballantyne, and I. M. Davis. "Core Stability Measures as Risk Factors for Lower Extremity Injury in Athletes." *Medicine & Science in Sports & Exercise* 36, no. 6 (2004): 926–934.

McHugh, M. P., and C. H. Cosgrave. "To Stretch or Not to Stretch: The Role of Stretching in Injury Prevention and Performance." *Scandinavian Journal of Medicine & Science in Sports* 20, no. 2 (2010): 169–181.

Nielsen, J., K. D. Gejl, M. Hey-Mogensen, H.-C. Holmberg, C. Suetta, P. Krustrup et al. "Plasticity in Mitochondrial Cristae Density Allows Metabolic Capacity Modulation in Human Skeletal Muscle." *Journal of Physiology* 595, no. 9 (2017); 2839–2847.

Nielsen, R. O., E. A. Nohr, S. Rasmussen, and H. Sørensen. "Classifying Running-Related Injuries Based upon Etiology, with Emphasis on Volume and Pace." *International Journal of Sports Physical Therapy* 8, no. 2 (2013): 172–179.

Nielsen, R. O., I. Buist, H. Sørensen, M. Lind, and S. Rasmussen. "Training Errors and Running Related Injuries: A Systematic Review." *Journal of Orthopaedic and Sports Physical Therapy* 7, no. 1 (2012): 58–75.

Saragiotto, B. T., T. P. Yamato, and A. D. Lopes. "What Do Recreational Runners Think about Risk Factors for Running Injuries? A Descriptive Study of Their Beliefs and Opinions." *Journal of Orthopaedic & Sports Physical Therapy* 44, no. 10 (2014): 733–738.

Sato, K., and M. Mokha. "Does Core Strength Training Influence Running Kinetics, Lower-Extremity Stability, and 5000-m Performance in Runners?" *Journal of Strength & Conditioning Research* 23, no. 1 (2009): 133–140.

Schubert, A. G., J. Kempf, and B. C. Heiderscheit. "Influence of Stride Frequency and Length on Running Mechanics: A Systematic Review." *Sports Health* 6, no. 3 (2014): 210–217.

Tenforde, A. S., and M. Fredericson. "Influence of Sports Participation on Bone Health in the Young Athlete: A Review of the Literature." *PM&R* 3, no. 9 (2011): 861–867.

Tessutti, V., A. P. Ribeiro, F. Trombini-Souza, and I. C. N. Sacco. "Attenuation of Foot Pressure during Running on Four Different Surfaces: Asphalt, Concrete, Rubber, and Natural Grass." *Journal of Sports Sciences* 30, no. 14 (2012): 1545–1550.

Van Gent, R. N., D. Siem, M. van Middelkoop, A. G. van Os, S. M. A. Bierma-Zeinstra, and B. W. Koes. "Incidence and Determinants of Lower Extremity Running Injuries in Long Distance Runners: A Systematic Review." *British Journal of Sports Medicine* 41, no. 8 (2007): 469–480.

Williams, D. S. "High-Arched Runners Exhibit Increased Leg Stiffness Compared to Low-Arched Runners." *Gait and Posture* 19, no. 3 (2004): 263–269.

Willy R. W. "Innovations and Pitfalls in the Use of Wearable Devices in the Prevention and Rehabilitation of Running Related Injuries." *Physical Therapy in Sport* 29 (2018): 26–33.

6   THE RIGHT FUEL IN THE TANK

Bellinger, P. M. "β-Alanine Supplementation for Athletic Performance: An Update." *Journal of Strength & Conditioning Research* 28, no. 6 (2014): 1751–1770.

Berg, J. M., J. L. Tymoczko, and L. Stryer. *Biochemistry.* 5th ed. New York: W. H. Freeman, 2002.

Burke, L. M., M. L. Ross, L. A. Garvican-Lewis, M. Welvaert, I. A. Heikura, S. G. Forbes, et al. "Low Carbohydrate, High Fat Diet Impairs Exercise Economy and Negates the Performance Benefit from Intensified Training in Elite Race Walkers." *Journal of Physiology* 595, no. 9 (2017): 2785–2807.

Cermak, N. M., M. J. Gibala, and L. J. C. van Loon. "Nitrate Supplementation's Improvement of 10-km Time-Trial Performance in Trained Cyclists." *International Journal of Sport Nutrition and Exercise Metabolism* 22, no. 1 (2012): 64–71.

Ganio, M. S., J. F. Klau, D. J. Casa, L. E. Armstrong, and C. M. Maresh. "Effect of Caffeine on Sport-Specific Endurance Performance: A Systematic Review." *Journal of Strength and Conditioning Research* 23, no. 1 (2009): 315–324.

Gerlach, K. E., H. W. Burton, J. M. Dorn, J. J. Leddy, and P. J. Horvath. "Fat Intake and Injury in Female Runners." *Journal of the International Society of Sports Nutrition* 5, no. 1 (2008).

Jeukendrup, A. E. "Multiple Transportable Carbohydrates and Their Benefits." *Sports Science Exchange*, 2013; 26 (108): 1–5

Jonvik, K. L., J. Nyakayiru, L. J. C. van Loon, and L. B. Verdijk. "Can Elite Athletes Benefit from Dietary Nitrate Supplementation?" *Journal of Applied Physiology* 119, no. 6 (2015): 759–761.

Kato, H., K. Suzuki, M. Bannai, and D. R. Moore. "Protein Requirements Are Elevated in Endurance Athletes after Exercise as Determined by the Indicator Amino Acid Oxidation Method." *PLOS ONE* 11, no. 6 (2016): e0157406.

Knab, A. M., R. A. Shanely, K. D. Corbin, F. Jin, W. Sha, and D. C. Nieman. "A 45-Minute Vigorous Exercise Bout Increased Metabolic Rate for 14 Hours." *Medicine & Science in Sports & Exercise* 43, no. 9 (2011): 1643–1648.

Mudde, T. "Wie veel bietensap drinkt 'speelt Russische roulette' met gezondheid." *De Volkskrant*, October 31, 2013.

Noakes, T. *Lore of Running.* Cape Town: Oxford University Press Southern Africa, 2001.

Noakes, T. D., J. S. V, and S. D. Phinney. "Low Carbohydrate Diets for Athletes: What Evidence?" *British Journal of Sports Medicine* 48, no. 14 (2014): 1077–1078.

Nyakayiru, J. M., K. L. Jonvik, P. J. M. Pinckaers, J. Senden, L. J. C. van Loon, and L. B. Verdijk. "No Effect of Acute and 6-Day Nitrate

Supplementation on VO_2 and Time-Trial Performance in Highly Trained Cyclists." *International Journal of Sport Nutrition and Exercise Metabolism* 27, no. 1 (2017): 11–17.

Pontzer, H., R. Durazo-Arvizu, L. R. Dugas, J. Plange-Rhule, P. Bovet, T. E. Forrester, et al. "Constrained Total Energy Expenditure and Metabolic Adaptation to Physical Activity in Adult Humans." *Current Biology* 26, no. 3 (2016): 410–417.

Rodriguez, N. R., N. M. DiMarco, S. Langley, et al. "Position of the American Dietetic Association, Dietitians of Canada, and the American College of Sports Medicine: Nutrition and Athletic Performance." *Journal of the American Dietetic Association* 109, no. 3 (2009): 509–527.

Shaw, K. A., H. C. Gennat, P. O'Rourke, and C. Del Mar. "Exercise for Overweight or Obesity." *Cochrane Database of Systematic Reviews* 18, no. 4 (2006): CD003817.

Volek, J. S., T. Noakes, and S. D. Phinney. "Rethinking Fat as a Fuel for Endurance Exercise." *Journal of European Journal of Sport Science* 15, no. 1 (2015): 13–20.

Volkers, J. "Beenblessure Sven Kramer was te wijten aan te veel vitamine B6." *De Volkskrant*, November 14, 2015.

Whitfield, J., A. Ludzki, G. J. F. Heigenhauser, J. M. G. Senden, L. B. Verdijk, L. J. C. van Loon, et al. "Beetroot Juice Supplementation Reduces Whole Body Oxygen Consumption but Does Not Improve Indices of Mitochondrial Efficiency in Human Skeletal Muscle." *Journal of Physiology* 594, no. 2 (2016): 421–435.

7 A SPRINT TO THE JOHN

Allen, J. M., L. J. Mailing, G. M. Niemiro, R. Moore, M. D. Cook, B. A. White, et al. "Exercise Alters Gut Microbiota Composition and Function in Lean and Obese Humans." *Medicine & Science in Sports & Exercise* 47, no. 1 (2017): 101–110.

Almond, C. S. D., A. Y. Shin, E. B. Fortescue, R. C. Mannix, D. Wypij, B. A. Binstadt, et al. "Hyponatremia among Runners in the Boston Marathon." *New England Journal of Medicine* 352, no. 15 (2005): 1550–1556.

Cohen, D. "The Truth about Sports Drinks." *British Medical Journal* 345 (2012): e4737.

De Ataide e Silva, T., M. E. de Souza, J. F. de Amorim, C. G. Stathis, C. G. Leandro, and Adriano Eduardo Lima-Silva. "Can Carbohydrate Mouth Rinse Improve Performance during Exercise? A Systematic Review." *Nutrients* 6, no. 1 (2014): 1–10.

FitBiomics. Understanding Elite Microbiomes for Performance and Recovery Applications. Abstract. Presented in August 2017 at a meeting of the American Chemical Society.

Goulet, E. D. "Effect of Exercise-Induced Dehydration on Time-Trial Exercise Performance: A Meta-Analysis." *British Journal of Sports Medicine* 45, no. 14 (2011): 1149–1156.

Jeukendrup, A. E. "Training the Gut for Athletes." *Sports Medicine* 47, no. 1 (2017): 101–110.

Mach, N., and D. Fuster-Botella. "Endurance Exercise and Gut Microbiota: A Review." *Journal of Sport and Health Science* 6, no. 2 (2017): 179–197.

Murray, R. "Training the Gut for Competition." *Current Sports Medicine Reports* 5, no. 3 (2006): 161–164.

Parr, E. B., D. M. Camera, J. L. Areta, L. M. Burke, S. M. Phillips, J. A. Hawley, and V. G. Coffey. "Alcohol Ingestion Impairs Maximal Post-Exercise Rates of Myofibrillar Protein Synthesis Following a Single Bout of Concurrent Training." *PLOS ONE* 9, no. 2 (2014): e88384.

Rehrer, N. J., G. M. Janssen, F. Brouns, and W. H. Saris. "Fluid Intake and Gastrointestinal Problems in Runners Competing in a 25-km Race and a Marathon." *International Journal of Sports Medicine* 10, no. 1 (1989): 22–25.

Rosner, M. H., and J. Kirven. "Exercise-Associated Hyponatremia." *Clinical Journal of the American Society of Nephrology* 2, no. 1 (2007): 151–161.

Simons, S. M., and G. G. Shaskan. "Gastrointestinal Problems in Distance Running." *International SportMed Journal* 6, no. 3 (2005): 162–170.

Ter Steege, R. W. F., J. Van Der Palen, and J. J. Kolkman. "Prevalence of Gastrointestinal Complaints in Runners Competing in a Long-Distance Run: An Internet-Based Observational Study in 1281 Subjects." *Scandinavian Journal of Gastroenterology* 43, no. 12 (2008): 1477–1482.

Wijnen, A. H. C., J. Steennis, M. Catoire, F. C. Wardenaar, and M. Mensink. "Post-Exercise Rehydration: Effect of Consumption of Beer with Varying Alcohol Content on Fluid Balance after Mild Dehydration." *Frontiers in Nutrition* 3, no. 45 (2016).

Wolin, K. Y., Y. Yan, G. A. Colditz, and I.-M. Lee. "Physical Activity and Colon Cancer Prevention: A Meta-Analysis." *British Journal of Cancer* 100, no. 4 (2009): 611–616.

8   RUNNING FOR YOUR LIFE

Aengevaeren, V. L., A. Mosterd, T. L. Braber, N. H. J. Prakken, P. A. Doevendans, D. E. Grobbee, et al. "Relationship between Lifelong Exercise Volume and Coronary Atherosclerosis in Athletes." *Circulation* 136, no. 2 (2017): 138–148.

Arem, H., S. C. Moore, A. Patel, P. Hartge, A. B. de Gonzalez, K. Visvanathan, et al. "Leisure Time Physical Activity and Mortality: A Detailed Pooled Analysis of the Dose-Response Relationship." *JAMA Internal Medicine* 175, no. 6 (2015): 959–967.

Bohm, P., G. Schneider, L. Linneweber, A. Rentzsch, N. Krämer, H. Abdul-Khaliq, et al. "Right and Left Ventricular Function and Mass in Male Elite Master Athletes: A Controlled Contrast-Enhanced Cardiovascular Magnetic Resonance Study." *Circulation* 133, no. 20 (2016): 1927–1935.

Diaz, K. M., V. J. Howard, B. Hutto, N. Colabianchi, J. E. Vena, M. M. Safford, et al. "Patterns of Sedentary Behavior and Mortality in U.S. Middle-Aged and Older Adults: A National Cohort Study." *Annals of Internal Medicine* 167, no. 7 (2017): 465–475.

Duvivier, B. M. F. M., N. C. Schaper, Michelle A. Bremers, G. van Crombrugge, P. P. C. A. Menheere, M. Kars, and H. H. C. M. Savelberg. "Minimal Intensity Physical Activity (Standing and Walking) of Longer Duration Improves Insulin Action and Plasma Lipids More

Than Shorter Periods of Moderate to Vigorous Exercise (Cycling) in Sedentary Subjects When Energy Expenditure Is Comparable." *PLOS ONE* 8, no. 2 (2013): e55542.

Eijsvogels, T., and P. D. Thompson. "Are There Clinical Cardiac Complications from Too Much Exercise?" *Current Sports Medicine Reports* 16, no. 1 (2017): 9–11.

Eijsvogels, T., and P. D. Thompson. "Exercise Is Medicine: At Any Dose?" *Journal of the American Medical Association* 314 (2015): 1915Y6.

Kim, J. H., R. Malhotra, G. Chiampas, P. d'Hemecourt, C. Troyanos, J. Cianca et al. "Cardiac Arrest during Long-Distance Running." *New England Journal of Medicine* 366 (2012): 130–140.

Kyu, H. H., V. F. Bachman, L. T. Alexander, J. E. Mumford, A. Afshin, K. Estep, et al. "Physical Activity and Risk of Breast Cancer, Colon Cancer, Diabetes, Ischemic Heart disease, and Ischemic Stroke Events: Systematic Review and Dose-Response Meta-Analysis for the Global Burden of Disease Study 2013." *British Medical Journal* 354 (2016: i3857.

Landry, C. H., K. S. Allan, K. A. Connelly, K. Cunningham, L. J. Morrison, P. Dorian, et al. "Sudden Cardiac Arrest during Participation in Competitive Sports." *New England Journal of Medicine* 377, no. 20 (2017): 1943–1953.

Lee, D., A. G. Brellenthin, P. D. Thompson, X. Sui, I-M. Lee, and C. J. Lavie. "Running as a Key Lifestyle Medicine for Longevity." *Progress in Cardiovascular Diseases* 60, no. 1 (2017): 45–55.

Lieberman, D. E. "Is Exercise Really Medicine? An Evolutionary Perspective." *Current Sports Medicine Reports* 14, no. 4 (2015): 313–318.

O'Donovan, G., I-M. Lee, M. Hamer, and E. Stamatakis. "Association of 'Weekend Warrior' and Other Leisure Time Physical Activity Patterns with Risks for All-Cause, Cardiovascular Disease, and Cancer Mortality." *JAMA Internal Medicine* 177, no. 3 (2017): 335–342.

O'Keefe, J. H., P. Schnohr, and C. J. Lavie. "The Dose of Running That Best Confers Longevity." *Heart* 99, no. 8 (2013): 588–590.

Raichlen, D. A., H. Pontzer, J. A. Harris, A. Z. P. Mabulla, F. W. Marlowe, J. J. Snodgrass, et al. "Physical Activity Patterns and Biomarkers of Cardiovascular Disease Risk in Hunter-Gatherers." *American Journal of Human Biology* 29, no. 2 (2017): e22919.

Schnohr, P., J. H. O'Keefe, J. L. Marott, P. Lange, and G. B. Jensen. "Dose of Jogging and Long-Term Mortality." *Journal of the American College of Cardiology* 65, no. 5 (2015): 411–419.

9   THE SECRET TO SPEED

Eclarinal, J. D., S. Zhu, M. S. Baker, D. B. Piyarathna, C. Coarfa, M. L. Fiorotto, and R. A. Waterland. "Maternal Exercise during Pregnancy Promotes Physical Activity in Adult Offspring." *FASEB Journal* 30, no. 7 (2016): 2541–2548.

Hoogkamer, W., R. Kram, and C. J. Arellano. "How Biomechanical Improvements in Running Economy Could Break the 2-Hour Marathon Barrier." *Sports Medicine* 47, no. 9 (2017): 1739–1750.

Huppertz, C. "How Voluntary Exercise Behavior Runs in Families: Twin Studies and Beyond." Doctoral thesis, Free University, 2016.

Joyner, M. J. "Modeling: Optimal Marathon Performance on the Basis of Physiological Factors." *Journal of Applied Physiology* 70, no. 2 (1991): 683–687.

Joyner, M. J., J. R. Ruiz, and A. Lucia. "The Two-Hour Marathon: Who and When?" *Journal of Applied Physiology* 110, no. 1 (2011): 275–277.

Noakes, T. D. "Fatigue Is a Brain-Derived Emotion That Regulates the Exercise Behavior to Ensure the Protection of Whole Body Homeostasis." *Frontiers in Physiology* 3, no. 82 (2012): 1–13.

Noakes, T. D. "Testing for Maximum Oxygen Consumption Has Produced a Brainless Model of Human Exercise Performance." *British Journal of Sports Medicine* 42, no. 7 (2008): 551–555.

Pitsiladis, Y. P., V. O. Onywera, E. Geogiades, W. O'Connell, and M. K. Boit. "The Dominance of Kenyans in Distance Running." *Equine and Comparative Exercise Physiology* 1, no. 4 (2004): 285–291.

Saltin, B., H. Larsen, N. Terrados, J. Bangsbo, T. Bak, C. K. Kim, et al. "Aerobic Exercise Capacity at Sea Level and at Altitude in Kenyan

Boys, Junior and Senior Runners Compared with Scandinavian Runners." *Scandinavian Journal of Medicine & Science in Sports* 5, no. 4 (1995): 209–221.

Saltin, B., C. K. Kim, N. Terrados, H. Larsen, J. Svedenhag, and C. J. Rolf. "Morphology, Enzyme Activities and Buffer Capacity in Leg Muscles of Kenyan and Scandinavian Runners." *Scandinavian Journal of Medicine & Science in Sports* 5, no. 4 (1995): 222–230.

Saunders, P. U., D. B. Pyne, R. D. Telford, and J. A. Hawley. "Factors Affecting Running Economy in Trained Distance Runners." *Sports Medicine* 34, no. 7 (2004): 465–485.

Tawa, N., and Q. Louw. "Biomechanical Factors Associated with Running Economy and Performance of Elite Kenyan Distance Runners: A Systematic Review." *Journal of Bodywork & Movement Therapies* 22, no. 1 (2018): 1–10.

Wilber, R. L., and Y. P. Pitsiladis. "Kenyan and Ethiopian Distance Runners: What Makes Them So Good?" *International Journal of Sports Physiology and Performance* 7, no. 2 (2012): 92–102.

Zavorsky, G. S., K. A. Tomko, and J. M. Smoliga. "Declines in Marathon Performance: Sex Differences in Elite and Recreational Athletes." *PLOS ONE* 12, no. 2 (2017): e0172121.

10 FATIGUE IS ALL IN THE MIND

Allen, E. J., P. M. Dechow, D. G. Pope, and George Wu. "Reference-Dependent Marathon Preferences: Evidence from Marathon Runners." NBER Working Paper No. 20343, July 2014.

Bacon, C., T. R. Myers, and C. I. Karageorghis. "Effect of Music-Movement Synchrony on Exercise Oxygen Consumption." *Journal of Physical Fitness and Sports Medicine* 52, no. 4 (2012): 359–365.

Blanchfield, A. W., J. Hardy, and S. Marcora. "Non-conscious Visual Cues Related to Affect and Action Alter Perception of Effort and Endurance Performance." *Frontiers in Human Neuroscience* 8 (2014): 967.

Doherty, M., and P. M. Smith. "Effects of Caffeine Ingestion on Rating of Perceived Exertion during and after Exercise: A Meta-Analysis."

Scandinavian Journal of Medicine & Science in Sports 15, no. 2 (2005): 69–78.

Fullagar, H. H. K., S. Skorski, R. Duffield, D. Hammes, A. J. Coutts, and T. Meyer. "Sleep and Athletic Performance: The Effects of Sleep Loss on Exercise Performance, and Physiological and Cognitive Responses to Exercise." *Sports Medicine* 45, no. 2 (2015): 161–186.

Karageorghis, C. I., and D.-L. Priest. "Music in Sport and Exercise: An Update on Research and Application." *Sport Journal* 11, no. 3 (2008).

Lahaye, R., T. Waanders, and V. Hutter. "Verkering in de wetenschap: Lichaam & brein!" *SportknowhowXL*, June 24, 2014.

Lane, A. M., P. A. Davis, and T. J. Devonport. "Effects of Music Interventions on Emotional States and Running Performance." *Journal of Sports Science and Medicine* 10, no. 2 (2011): 400–407.

Marcora, S. M., W. Staiano, and V. Manning. "Mental Fatigue Impairs Physical Performance in Humans." *Journal of Applied Physiology* 106, no. 3 (2009): 857–864.

Marcora, S. M., and W. Staiano. "The Limit to Exercise Tolerance in Humans: Mind over Muscle?" *European Journal of Applied Physiology* 109, no. 4 (2010): 763–770.

McCormick, A., C. Meijen, and S. Marcora. "Psychological Determinants of Whole-Body Endurance Performance." *Sports Medicine* 45, no. 7 (2015): 997–1015.

Pageaux, B., S. M. Marcora, V. Rozand, and R. Lepers. "Mental Fatigue Induced by Prolonged Self-Regulation Does Not Exacerbate Central Fatigue during Subsequent Whole-Body Endurance Exercise." *Frontiers in Human Neuroscience* 9, no. 67 (2015).

Pageaux, B. "Perception of Effort in Exercise Science: Definition, Measurement and Perspectives." *European Journal of Sport Science* 16, no. 8 (2016): 885–894.

Samson A., D. Simpson, C. Kamphoff, and A. Langlier. "Think Aloud: An Examination of Distance Runners' Thought Processes." *International Journal of Sport and Exercise Psychology* 15, no. 2 (2015): 1–14.

Tenenbaum, G., R Lidor, N. Lavyan, K. Morrow, S. Tonnel, A. Gershgoren, et al. "The Effect of Music Type on Running Perseverance and Coping with Effort Sensations." *Psychology of Sport and Exercise* 5, no. 2 (2004): 89–109.

Terry, P. C., C. I. Karageorghis, A. M. Saha, and S. D'Auria. "Effects of Synchronous Music on Treadmill Running among Elite Triathletes." *Journal of Science and Medicine in Sport* 15, no. 1 (2012): 52–57.

Thomas, K., M. Elmeua, G. Howatson, and S. Goodall. "Intensity-dependent Contribution of Neuromuscular Fatigue after Constant-Load Cycling." *Medicine and Science in Sports and Exercise* 48, no. 9 (2016): 1751–1760.

Van Cutsem, J., S. Marcora, K. De Pauw, S. Bailey, R. Meeusen, and B. Roelands. "The Effects of Mental Fatigue on Physical Performance: A Systematic Review." *Sports Medicine* 47, no. 8 (2017): 1–20.

11 RUNNING AS THERAPY FOR THE BRAIN

Bąbel, P. "Memory of Pain Induced by Physical Exercise." *Memory* 24, no. 4 (2016): 548–559.

Basso, J. C., and W. A. Suzuki. "The Effects of Acute Exercise on Mood, Cognition, Neurophysiology, and Neurochemical Pathways: A Review." *Brain Plasticity* 2, no. 2 (2017): 127–152.

Berczik, K., A. Szabó, M. D. Griffiths, T. Kurimay, B. Kun, R. Urbán, and Z. Demetrovics. "Exercise Addiction: Symptoms, Diagnosis, Epidemiology, and Etiology." *Substance Use & Misuse* 47, no. 4 (2012): 403–417.

Boecker, H., T. Sprenger, M. E. Spilker, G. Henriksen, M. Koppenhoefer, K. J. Wagner, et al. "The Runner's High: Opioidergic Mechanisms in the Human Brain." *Cerebral Cortex* 18, no. 11 (2008): 2523–2531.

Bossers, W. J. R., L. H. V. van der Woude, F. Boersma, T. Hortobágyi, E. J. A. Scherder, and M. J. G. van Heuvelen. "Comparison of Effect of Two Exercise Programs on Activities of Daily Living in Individuals with Dementia: A 9-Week Randomized, Controlled Trial." *Journal of the American Geriatrics Society* 64, no. 6 (2016): 1258–1266.

Bossers, W. J. R., L. H. V. van der Woude, F. Boersma, T. Hortobágyi, E. J. A. Scherder, and M. J. G. van Heuvelen. "A 9-Week Aerobic and Strength Training Program Improves Cognitive and Motor Function in Patients with Dementia: A Randomized, Controlled Trial." *American Journal of Geriatric Psychiatry* 23, no. 11 (2015): 1106–1116.

Cooney, G. M., K. Dwan, C. A. Greig, D. A. Lawlor, J. Rimer, F. R. Waugh, et al. "Exercise for Depression." *Cochrane Database Systematic Reviews* 12, no. 9 (2013): CD004366.

Danielsson, L., A. M. Noras, M. Waern, and J. Carlsson. "Exercise in the Treatment of Major Depression: A Systematic Review Grading the Quality of Evidence." *Physiotherapy Theory and Practice* 29, no. 8 (2013): 573–585.

De Vries, J. D., M. L. M. van Hooff, S. A. E. Geurts, and M. A. J. Kompier. "Exercise to Reduce Work-Related Fatigue among Employees: A Randomized Controlled Trial." *Scandinavian Journal of Work, Environment & Health* 43, no. 4 (2017): 337–349.

De Vries, J. D., M. L. M. van Hooff, S. A. E. Geurts, and M. A. J. Kompier. "Exercise as an Intervention to Reduce Study-Related Fatigue among University Students: A Two-Arm Parallel Randomized Controlled Trial." *PLOS ONE* 11, no. 3 (2016): e0152137.

Eich, T. S., and J. Metcalfe. "Effects of the Stress of Marathon Running on Implicit and Explicit Memory." *Psychonomic Bulletin & Review* 16, no. 3 (2009): 475–479.

Erickson, K. I., M. W. Voss, R. S. Prakash, C. Basak, A. Szabo, L. Chaddock, et al. "Exercise Training Increases Size of Hippocampus and Improves Memory." *Proceedings of the National Academy of Sciences* 108, no. 7 (2011): 3017–3022.

Freimuth, M., S. Moniz, and S. R. Kim. "Clarifying Exercise Addiction: Differential Diagnosis, Co-Occurring Disorders, and Phases of Addiction." *International Journal of Environmental Research and Public Health* 8, no. 10 (2011): 4069–4081.

Fuss, J., J. Steinle, L. Bindila, M. K. Auer, H. Kirchherr, B. Lutz, and P. Gass. "A Runner's High Depends on Cannabinoid Receptors in Mice." *Proceedings of the National Academy of Sciences* 112, no. 42 (2015): 13105–13108.

Griffiths, M. "Exercise Addiction: A Case Study." *Addiction Research* 5, no. 2 (1997): 161–168.

Hausenblas, H. A., K. Schreiber, and J. M. Smoliga. "Addiction to Exercise." *BMJ* 357 (2017): j1745.

Kim E. J., B. Pellman, and J. J. Kim. "Stress Effects on the Hippocampus: A Critical Review." *Learning & Memory* 22, no. 9 (2015): 411–416.

Krogh, J., M. Nordentoft, J. A. Sterne, and D. A. Lawlor. "The Effect of Exercise in Clinically Depressed Adults: Systematic Review and Meta-Analysis of Randomized Controlled Trials." *Journal of Clinical Psychiatry* 72, no. 4 (2010): 529–538.

Kruisdijk, F. R., M. Hopman-Rock, A. T. F. Beekman, and I. Hendriksen. "EFFORT-D: Results of a Randomised Controlled Trial Testing the EFFect of Running Therapy on Depression." *BMC Psychiatry* 19, no. 170 (2019).

Northey, J. M., N. Cherbuin, K. L. Pumpa, D. J. Smee, and B. Rattray. "Exercise Interventions for Cognitive Function in Adults Older Than 50: A Systematic Review with Meta-Analysis." *British Journal of Sports Medicine* 298, no. 2 (2018): 372–377.

Potenza, M. N. "Non-Substance Addictive Behaviors in the Context of DSM-5." *Addictive Behaviors* 39, no. 1 (2014): 1–2.

Schuch, F. B., D. Vancampfort, J. Richards, S. Rosenbaum, P. B. Ward, and B. Stubbs. "Exercise as a Treatment for Depression: A Meta-Analysis Adjusting for Publication Bias." *Journal of Psychiatric Research* 77 (2016): 42–51.

Seifert T., P. Brassard, M. Wissenberg, P. Rasmussen, P. Nordby, B. Stallknecht, et al. "Endurance Training Enhances BDNF Release from the Human Brain." *American Journal of Physiology—Regulatory, Integrative and Comparative Physiology* 298, no. 2 (2010): 372–377.

Thomas, S. "Nowhere to Run: Did My Fitness Addiction Make Me Ill?" *Guardian*, March 7, 2015.

Thompson, C. J., K. Boddy, K. Stein, R. Whear, J. Barton, and M. H. Depledge. "Does Participating in Physical Activity in Outdoor Natural Environments Have a Greater Effect on Physical and Mental

Wellbeing Than Physical Activity Indoors? A Systematic Review." *Environmental Science & Technology* 45, no. 5 (2011): 1761–1772.

Wrann, C. D., J. P. White, J. Salogiannis, D. Laznik-Bogoslavski, J. Wu, D. Ma, et al. "Exercise Induces Hippocampal BDNF through a PGC-1α/FNDC5 Pathway." *Cell Metabolism* 18, no. 5 (2013): 649–659.

EPILOGUE

Keulemans, M. "Magische strohalmen in de sport." *De Volkskrant*, October 8, 2012. https://www.volkskrant.nl/nieuws-achtergrond/magische -strohalmen-in-de-sport~b3c4df33/.

INDEX

ACE, 83–87, 193

Achilles tendon, 5, 13, 56, 62, 86, 96, 193
  injury, 61, 99–100, 111

Active behavior, 191

ACTN3, 81–83, 85–87, 193

Adaptation, 98, 163

Alcohol, 163–165, 167, 176

Anandamide, 238–240

Ancestors, 5, 7, 38, 42–43, 48, 170, 192
  injuries, 9–12, 15–16

Anemia, 147, 152

Anti-pronation shoes, 34–36, 45, 263

Arch (foot), 13, 24, 36, 44, 46–47, 56, 58, 109

Artery calcification, 179–180, 185

Arthritis, 105–107, 184

Athlete's heart, 176–177, 179

Athletic talent, 80, 156, 191, 194, 196, 200–202, 214

ATP, 124–127

Bare feet, 41, 48–51, 59, 63–64

Barefoot running, 41–42, 48–54, 58–59, 63

BDNF, 244–246

Beetroot juice, 123, 142–146, 259

Bicarbonate, 146–147

Big data, 103

Bikila, Abebe, 50–51

Blade runner, 13–14

BMI, 29, 74

Bolt, Usain, 8, 101, 200

Bone density, 97–98, 175

Brain (healthy), 235, 240–246

Brain training, 228, 234

Breasts, 69–72, 78

Budd, Zola, 50–51

Burnout, 236, 246–248, 252–253

Caffeine, 141–142, 260–261
Carbohydrates, 123, 129–131
    133–136, 148, 152, 166
  concentrated, 140, 162
  digestion, 154, 156
Carbon fiber plate, 204
Cartilage, 169, 174
Comfort filter, 261, 262
Compression socks, 113–115,
    121, 259, 262
Core stability, 47, 111–112

Dehydration, 158–159, 161–162
Dementia, 242–245
Depression, 235–236, 238, 244,
    246–247, 249–253
Diabetes, 169, 171, 172, 175
DNA test, 86, 92, 197
Drinking, 158–159, 161–165, 182
Dynamic stretching, 115–117

Elastic energy, 5, 13, 109, 117
Endorphins, 237–239
Energy expenditure, 110, 137–
    138, 226
Evolution, 4–5, 11–12, 15–16,
    18, 170, 225
Exercise addiction, 255
Exercise paradox, 170
Exercise test, 194–196
Exhaustion, 218–219, 221–222,
    247, 249, 254

Fatty acids, 79, 126–127, 135
Female body, 68
Fit (being), 2–3, 87, 133, 171,
    174, 176–177, 185, 188, 190,
    240, 245

Flat feet, 72–74
Foam (sole), 204
Foam roller, 118–119
Foot type, 29, 33–34, 261
Forefoot landing, 51–55
Fun run, 17

Gels, 123, 134, 137, 139–142,
    152, 166
Genetic makeup, 86, 88, 190–
    191, 196, 205
Gibb, Roberta, 68–69
Glycogen, 125–127, 133–134,
    163, 206
Goals, setting, 214–215, 217

Heart attack, 178, 180–181,
    183
Heart disease, 167, 169–170,
    174–175, 180, 184–185
Heel landing, 48, 51–55, 57–58,
    64 107
Hippocampus, 241, 244–245
Homo erectus, 5–8
Hunter-gatherer, 7, 10, 16, 167
Hypertrophic cardiomyopathy,
    182, 186
Hyponatremia, 160, 182

Impact shock, 23–24, 28, 36, 46,
    50, 111
Injury numbers, 21
Injury prone, 11–12, 20–21, 27,
    46, 73, 75, 102, 131
Injury risk, 24, 28, 29, 31–32,
    102, 111, 176

Jogging, rise of, 2, 158

Kenyans, 189, 192–193, 197
Ketones, 135
Kimetto, Dennis, 101, 204
Kipchoge, Eliud, 89, 203–206
Knee injury, 19–20, 72, 101

Lactate, 125–126, 147, 157, 164, 195–196
Lactic acid, 125, 147, 157, 206
Lieberman, Daniel, 4–10, 43, 48–49, 55, 59, 67, 170, 226
Life span, 58, 171, 173–174, 187
Load capacity, 93, 102–103
Long-distance running, 4, 16, 86, 91, 116–118, 192, 194
    fuel, 126, 134, 139
    stomach problems, 150, 160–161
    heart damage, 178, 183
Losing weight, 4, 136–138

Marathon runners, 80, 84–86, 88–89, 97, 152, 158–161, 181, 183, 142, 212, 215, 242
Maximal oxygen uptake (VO$_2$max), 194–198, 206–207
Maximalist running shoes, 60–62
Meat, 7, 128, 130, 139
Memory, 229, 241, 243–245
Mental fatigue, 223–225
Mental health problems, 234–235, 246–247, 251, 254–255
Microbiome, 155–157
Minimalist running shoes, 54, 56–57, 59, 62–63
Mitochondria, 79, 97, 125–126, 143

Motion control shoes, 24–26, 28, 30, 34–36, 38, 44–46
Motivation, 3, 199, 207, 211, 215–217, 219–220, 224, 229, 231
Muscle control, 96, 108, 113, 116, 118
Muscle fiber, 9, 78–80, 82, 84–85, 89–91, 96–97, 125–126, 193
Muscle recovery, 130, 139, 164, 263
Music, 229–234

Nervous system, 96, 113, 116, 156
Nitrate, 143–146

Overuse injuries, 11, 60, 73, 91, 93–95, 99–101, 103
Overweight, 21, 67, 74–75

Perception of effort, 221–227, 232–233
Perseverance, 211, 214
Persistence hunting, 6–7, 16
Pistorius, Oscar, 14
Physique, 67–68, 76–78, 88–89, 91–92, 193, 198, 200, 207, 213
    in relation to injuries, 73, 75, 102, 190
Placebo effect, 114–115, 238
Preferred movement path, 261
Premature mortality, 170, 174
Preventive medical checkup, 183–184
Pronation, 24, 26, 37

Protein (nutrient), 123, 128,
    130–131, 139–140, 148, 154,
    156, 165–166
Protein powder, 139
Pseudoscience, 263

Radcliffe, Paula, 149–150, 231
Recovery meal, 130, 148
Resistance training, 90, 96
Reward system, 163
Runner's high, 236–240
Running biomechanics, 44, 61,
    76
Running economy, 50, 55, 76–
    77, 110, 117, 193, 197–198,
    207
Running surface, 23, 39, 107–
    109, 191
Running technique, 10, 23, 49,
    52–53, 55–56, 64, 72, 198,
    201, 207, 259–260
  changing, 59, 61, 120

Scar tissue, 178
Schippers, Dafne, 202
Screening, 185–186
Self-regulation, 212–215
Sensors (technology), 35, 53, 103
Shin splints, 20, 57, 59, 111, 114
Shock absorption, 30, 34, 36
Sitting, 10, 98, 106, 137, 171–
    173, 176, 223
Sleep, 102, 141, 176, 228–229,
    234, 259
Sport genes, 86
Sports bra, 70–71
Sports drinks, 158, 159, 162

Sports nutrition, 123–124, 128,
    130, 142, 162
Sprained ankle, 11, 15, 100
Sprinting, 8–9, 82, 85, 91, 117,
    125, 147, 202
Stamina, 9, 79, 82–84, 86, 190,
    194–195, 201
  build up, 56, 84, 97, 176, 210,
    228
Static stretching, 116–118
Stomach problems, 131, 134,
    141, 150–153, 161, 163
Stress fracture, 11, 30–31, 57,
    120
Stride frequency, 104, 109–111
Sub2 marathon, 202–206
Supplements, 123, 129, 139–140,
    144–148, 157, 263
Sweat glands, 6–7

Tarahumara, 15–16, 51
Technology (in relation to
    injuries), 103, 104

Ultramarathon, 168, 176–176,
    180, 187–188, 211

Vaporfly, 204, 205
Vitamin C, 148
Vitamin pills, 146–148

Weekend warrior, 175–176